NATIONAL ACADEMIES Sciences Engineering Medicine

NATIONAL ACADEMIES PRESS
Washington, DC

Innovating Scholarship of Teaching and Learning in Health Professions Education

Convened July 16, 2024, October 23–24, 2024, and October 31, 2024

Erin Hammers Forstag, Patricia Cuff, and Erika Lin Chow,
Rapporteurs

Global Forum on Innovation in Health Professional Education

Board on Global Health

Health and Medicine Division

Proceedings of a Workshop

NATIONAL ACADEMIES PRESS 500 Fifth Street, NW Washington, DC 20001

This activity was supported by contracts between the National Academy of Sciences and Academic Collaborative for Integrative Health; Academy of Nutrition and Dietetics; Accreditation Council for Graduate Medical Education; American Academy of Nursing; American Association of Colleges of Osteopathic Medicine; American Board of Family Medicine; American Council of Academic Physical Therapy; American Dental Education Association; American Medical Association; American Nurses Credentialing Center; American Occupational Therapy Association; American Physical Therapy Association; American Psychological Association; American Speech-Language-Hearing Association; Association of American Medical Colleges; Association of Schools and Colleges of Optometry; Athletic Training Strategic Alliance; Columbia University Vagelos College of Physicians and Surgeons; Council on Social Work Education; Intealth; Massachusetts General Hospital Institute of Health Professions; National Academies of Practice; National Board for Certified Counselors, Inc. and Affiliates; National Board of Medical Examiners; National Council of State Boards of Nursing, Inc.; National League for Nursing; National Register of Health Service Psychologists; Physician Assistant Education Association; Society for Simulation in Healthcare; Texas Tech University Health Sciences Center; Uniformed Services University; University of California, San Francisco; U.S. Department of Veterans Affairs; and Weill Cornell Medicine—Qatar. Any opinions, findings, conclusions, or recommendations expressed in this publication do not necessarily reflect the views of any organization or agency that provided support for the project.

International Standard Book Number-13: 978-0-309-09902-8
International Standard Book Number-10: 0-309-09902-1
Digital Object Identifier: https://doi.org/10.17226/29089

This publication is available from the National Academies Press, 500 Fifth Street, NW, Keck 360, Washington, DC 20001; (800) 624-6242 or (202) 334-3313; http://www.nap.edu.

Printed in the United States of America.

Suggested citation: National Academies of Sciences, Engineering, and Medicine. 2025. *Innovating scholarship of teaching and learning in health professions education: Proceedings of a workshop*. Washington, DC: National Academies Press. https://doi.org/10.17226/29089.

The **National Academy of Sciences** was established in 1863 by an Act of Congress, signed by President Lincoln, as a private, nongovernmental institution to advise the nation on issues related to science and technology. Members are elected by their peers for outstanding contributions to research. Dr. Marcia McNutt is president.

The **National Academy of Engineering** was established in 1964 under the charter of the National Academy of Sciences to bring the practices of engineering to advising the nation. Members are elected by their peers for extraordinary contributions to engineering. Dr. John L. Anderson is president.

The **National Academy of Medicine** (formerly the Institute of Medicine) was established in 1970 under the charter of the National Academy of Sciences to advise the nation on medical and health issues. Members are elected by their peers for distinguished contributions to medicine and health. Dr. Victor J. Dzau is president.

The three Academies work together as the **National Academies of Sciences, Engineering, and Medicine** to provide independent, objective analysis and advice to the nation and conduct other activities to solve complex problems and inform public policy decisions. The National Academies also encourage education and research, recognize outstanding contributions to knowledge, and increase public understanding in matters of science, engineering, and medicine.

Learn more about the National Academies of Sciences, Engineering, and Medicine at **www.nationalacademies.org**.

Consensus Study Reports published by the National Academies of Sciences, Engineering, and Medicine document the evidence-based consensus on the study's statement of task by an authoring committee of experts. Reports typically include findings, conclusions, and recommendations based on information gathered by the committee and the committee's deliberations. Each report has been subjected to a rigorous and independent peer-review process and it represents the position of the National Academies on the statement of task.

Proceedings published by the National Academies of Sciences, Engineering, and Medicine chronicle the presentations and discussions at a workshop, symposium, or other event convened by the National Academies. The statements and opinions contained in proceedings are those of the participants and are not endorsed by other participants, the planning committee, or the National Academies.

Rapid Expert Consultations published by the National Academies of Sciences, Engineering, and Medicine are authored by subject-matter experts on narrowly focused topics that can be supported by a body of evidence. The discussions contained in rapid expert consultations are considered those of the authors and do not contain policy recommendations. Rapid expert consultations are reviewed by the institution before release.

For information about other products and activities of the National Academies, please visit www.nationalacademies.org/about/whatwedo.

WORKSHOP PLANNING COMMITTEE

KIM DUNLEAVY (*Cochair*), University of Florida
PAMELA JEFFRIES (*Cochair*), Vanderbilt University
JONATHAN AMIEL, New York-Presbyterian Hospital
GRACE HUANG, Harvard Medical School
GAIL JENSEN, Creighton University
LORETTA NUNEZ, American Speech-Language-Hearing Association
YASUHARU OKUDA, University of South Florida Health Center for Advanced Medical Learning and Simulation

Staff

PATRICIA CUFF, Director, Global Forum on Innovation in Health Professional Education
ERIKA LIN CHOW, Research Associate
TAYLOR WINDMILLER, Senior Program Assistant
JULIE PAVLIN, Science Advisor, Health and Medicine Division; Board Director, Board on Global Health

Reviewers

This Proceedings of a Workshop was reviewed in draft form by individuals chosen for their diverse perspectives and technical expertise. The purpose of this independent review is to provide candid and critical comments that will assist the National Academies of Sciences, Engineering, and Medicine in making each published proceedings as sound as possible and to ensure that it meets the institutional standards for quality, objectivity, evidence, and responsiveness to the charge. The review comments and draft manuscript remain confidential to protect the integrity of the process.

We thank the following individuals for their review of this proceedings:

SEAN BRUNA, Agency for Healthcare Research and Quality
HILLIARY E. CREELY, Indiana University of Pennsylvania
HANNA PHAN, University of Michigan

Although the reviewers listed above provided many constructive comments and suggestions, they were not asked to endorse the content of the proceedings, nor did they see the final draft before its release. The review of this proceedings was overseen by **HUGH TILSON,** the University of North Carolina at Chapel Hill. He was responsible for making certain that an independent examination of this proceedings was carried out in accordance with standards of the National Academies and that all review comments were carefully considered. Responsibility for the final content rests entirely with the rapporteurs and the National Academies.

Acknowledgments

The National Academies of Sciences, Engineering, and Medicine's Global Forum on Innovation in Health Professional Education wishes to express its sincere gratitude to focus group participants for volunteering their time at information-gathering sessions during the workshop planning phase. The forum also expresses deep appreciation to the published authors who provided video recordings presenting their evidence-based Scholarship of Teaching and Learning research. Lastly, the forum wishes to thank all speakers and planning committee members, who collaborated to ensure a workshop replete with informative presentations and rich moderated discussions.

Contents

APPENDIXES

Boxes and Figures

BOXES

FIGURES

1

Introduction

On October 23 and 24, 2024, the Global Forum on Innovation in Health Professional Education of the National Academies of Sciences, Engineering, and Medicine (the forum) held a workshop on the Scholarship of Teaching and Learning in Health Professions Education. The workshop series was planned by a committee according to a statement of task (see Box 1-1) and included several components over a 4-month period (see Figure 1-1). The planning committee[1] used input from previous activities and virtual input to inform the planning of activities. First, a preworkshop session was held on July 16, 2024, to lay a foundation for a shared understanding of Scholarship of Teaching and Learning (SOTL) by reviewing the history of and terms used for SOTL. At this session, a roundtable discussion was held to look at how SOTL is applied in health professional education (HPE) and explore different priorities and perspectives on SOTL. In August, information-gathering sessions (focus groups) were held to inform planning committee members about the SOTL landscape, including individual and institutional perceptions, beliefs, and attitudes about SOTL within various health professions. In September, the planning committee invited authors to share publications that demonstrate how SOTL has impacted health

[1] The planning committee's role was limited to planning the workshop, and the Proceedings of a Workshop was prepared by the workshop rapporteurs as a factual summary of what occurred at the workshop. Statements, recommendations, and opinions expressed are those of individual presenters and participants and are not necessarily endorsed or verified by the National Academies of Sciences, Engineering, and Medicine, and they should not be construed as reflecting any group consensus.

BOX 1-1
Statement of Task

A planning committee of the National Academies of Sciences, Engineering, and Medicine will organize a public workshop to explore the systematic study of teaching and learning and the current or potential application of the Scholarship of Teaching and Learning (SOTL) in health professions education. Invited presentations and discussions will involve global audiences that will consider the use of SOTL in areas such as

- Recognizing and supporting SOTL for professional development, including academic promotion and tenure within health professions education;
- Identifying structures that motivate clinical educators and those involved in continuous professional development to attain training and competence in SOTL;
- Using SOTL to support innovations and adjustments required for the changing academic environment and learner profiles such as competency-based education, artificial intelligence (AI), admissions needs;
- Demonstrating health professions education impacts beyond satisfaction surveys; and
- Exploring the science of teaching and learning.

professions in key areas. On October 23 and 24, the main workshop was held, with the goals of amplifying the value of SOTL in HPE, exploring innovations in educational research frameworks, and considering what is important for the future of teaching practice in HPE. Last, a 1-hour closing session was held on October 31; planning committee members convened to discuss what was learned at the workshop and consider motivational messages that might entice more educators to engage in SOTL.

Chapter 1 of this proceedings describes the planning methodology and objectives of the workshop, and Chapter 2 summarizes the discussions held at the preworkshop session. Chapter 3 delves into the main workshop, with presentations on the value of SOTL in HPE and innovations in SOTL, and includes reports from breakout activities. Chapter 4 summarizes the final closing session, which featured an exploration of mechanisms for individual faculty development and a discussion of how to motivate more educators to engage in SOTL. Appendixes A, B, and C contain a list of references, the breakout group figures, and the full workshop agenda, respectively. Appendix D provides links to the evidence-based research videos that authors were invited to submit before the workshop. Biographies of the workshop planning committee members and speakers are provided in Appendixes E and F, respectively. Members of the forum that hosted the workshop are listed in Appendix G.

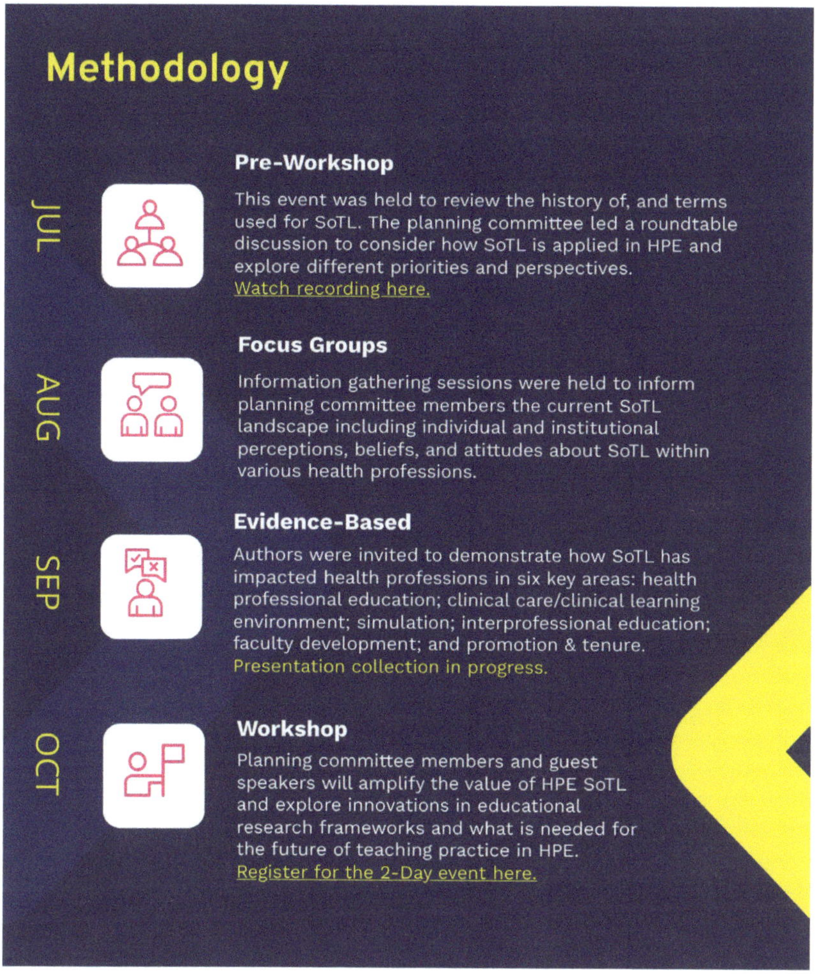

FIGURE 1-1 Methodology of the workshop.
NOTES: HPE = Health Professions Education; SoTL = Scholarship of Teaching and Learning.
SOURCE: Presented by Kim Dunleavy, October 31, 2024.

2

Preworkshop Session

The July 16, 2024, introductory session was opened by Jonathan Amiel, director of the Office of Professionalism and Inclusion in the Learning Environment at New York-Presbyterian and professor of psychiatry, Columbia University's Vagelos College of Physicians & Surgeons. The goals, said Amiel, were to review the history of and terms used in the Scholarship of Teaching and Learning (SOTL)[1] and lay a foundation for a shared understanding of it and how it is applied in health professional education (HPE). The session featured a speaker who presented the general history of traditional SOTL, followed by a roundtable discussion in which panelists delved into its practical applications in HPE. Amiel emphasized that the goal of the workshop series was to advance health professional workforce development to improve patient outcomes.

Amiel briefly introduced participants to the origin of SOTL. The traditional model of academic scholarship, he said, has long been focused almost exclusively on basic and clinical research that is peer reviewed and published in major journals. In 1990, Ernest Boyer proposed that this concept be broadened to include other valuable aspects of work; he argued that scholarship should include four categories:

---

[1]According to Jensen in the preworkshop, SOTL can be defined as evidence of impactful teaching that is interwoven with evidence of student learning. Importantly, it is the educator's inquiry—and learning by the educator and learner—that drive the process, and making the results public for others to review further informs the teaching and learning content and process.

- Scholarship of discovery,
- Scholarship of integration,
- Scholarship of application, and
- Scholarship of teaching and learning.

SOTL IN HPE

Gail Jensen, vice provost for learning and assessment, dean emerita, and professor of physical therapy, Creighton University, began by sharing Shulman's (2001) description of scholarship: "We cannot only count and name in scholarship; we can recount and teach ... in recounting, we transform and connect to integrate new forms of creations what might otherwise have been experienced individually." Shulman often said that teaching is like dry ice, said Jensen, because it "evaporates the second you stop." Due to this ephemeral nature, Jensen said, it is necessary to codify the connection between teaching and learning by recounting and integrating the process of teaching. Boyer (1990) laid a foundation for SOTL by pushing for a broader definition of scholarship that moved beyond the "teaching versus research" dichotomy, said Jensen, and other scholars built on this work. Glassick (2000) added the importance of standards for scholarship, and Shulman (2001) emphasized the need for inquiry related to student learning and for making the evidence of student learning publicly available. The scholarship of teaching, said Jensen, is a condition for excellent teaching (Hutchings and Shulman, 1999) and facilitates the traditional process of public peer review and critique by the community. In turn, she added, SOTL leads to scholarly teaching that is grounded in evidence, arguments, methods, and values (see Figure 2-1), so there is an iterative relationship between SOTL and scholarly teaching, with each impacting the other.

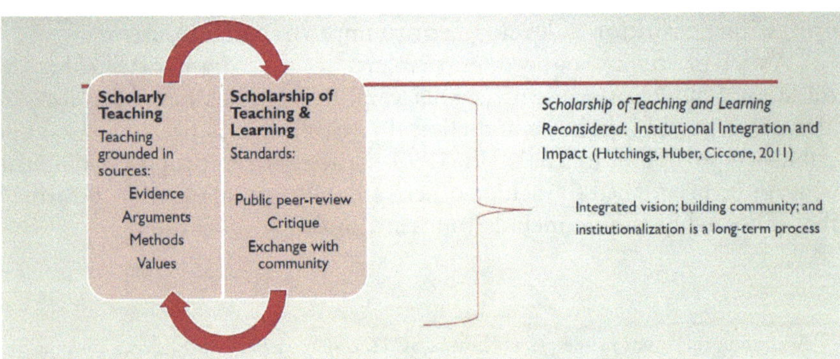

FIGURE 2-1 Relationship between SOTL and scholarly teaching.
NOTE: SOTL = Scholarship of Teaching and Learning.
SOURCES: Presented by Gail Jensen on July 16, 2024; Hutchings and Shulman, 1999; Hutchings et al., 2011.

The world of HPE has a natural connection between education and practice, said Jensen. Academic and clinical communities are connected through performance-based learning and assessment. Approaches such as practice-based learning and competency-based education provide opportunities for educators and learners to engage in education while reflecting on the process and outcomes. For health professionals, the ability to make judgments in uncertain conditions is critical, said Jensen, and learning in a clinical setting allows them to think critically and practice this skill. Several unique characteristics in HPE present opportunities to use SOTL to improve education and patient care: HPE is heavily grounded in performance-based learning and assessment, the faculty are a diverse group of educators and practitioners, and learning often occurs in settings of interprofessional collaborative practice. Furthermore, said Jensen, forming professional identity is critically important for HPE learners, as is the ability to bring "head and hand and heart" together when caring for patients. HPE has rich opportunities to use SOTL to move professions forward and drive change, said Jensen; however, supporting scholarship in HPE will require funding and collaboration among educators and their leadership.

ROUNDTABLE DISCUSSION

Amiel led a roundtable discussion with panelists representing different areas of HPE. He began by asking each one to describe what successful HPE SOTL would look like from their perspective. Pamela Jeffries, dean of the Vanderbilt University School of Nursing and planning committee cochair, said that it would result in both improved learning outcomes and increased knowledge about what teaching approaches lead to them. She noted that such improvements could be measured in a number of ways, from readiness for transition to practice to patient outcomes. Grace C. Huang, dean for faculty affairs and professor of medicine, Harvard Medical School, said that just as educational research is important, educational innovations are the "bread and butter" of SOTL. HPE has had a number of notable innovations in recent years, including simulation-based classrooms, implicit bias training, and trauma-informed care. Huang noted that when authors are seeking to publish their innovative ideas in a journal, their manuscripts are much stronger if their ideas are grounded in conceptual frameworks and theories. Loretta M. Nunez, senior director of academic affairs and research education, American Speech-Language-Hearing Association, said that successful SOTL provides the evidence for educational practices that address issues in HPE. For example, HPE has embraced evidence-based practices, such as competency-based education, culturally responsive teaching, and interprofessional education (IPE). Demonstrating the value of these practices through SOTL can be encouraged, funded, disseminated, and acknowledged as a valuable contribution to HPE, she said. Yasuharu

(Haru) Okuda, forum representative for the University of South Florida Health Center for Advanced Medical Learning and Simulation, spoke last and said that while SOTL is aimed at education and training, the ultimate goal can be to use its lessons to implement evidence-based education that improves patient safety and clinical outcomes.

Benefits of SOTL

Amiel noted the competing priorities in hospitals and health systems, and HPE and training are not always the highest. However, he said, prioritizing SOTL can enhance and amplify the impact of innovations on patient outcomes. Amiel asked panelists to share examples of how SOTL has led to meaningful change in patient outcomes or care. Okuda agreed that education and training are often lower priorities and can be the "first thing to go when times are tough"; education and training are expensive, and leaders may question the return on investment (ROI). Okuda said that in response to these types of concerns, he often shares an example about simulation-based training and its impact on health care costs: Barsuk and colleagues (2009) demonstrated that deliberate practice and mastery learning improved learning and clinical outcomes. Multiple studies on deliberate practice and mastery learning in the area of central line training showed that deliberate practice improved clinical outcomes and reduced infections and other complications (Allen et al., 2014; Barsuk et al., 2014). In one study, researchers estimated the cost savings related to reduced infections after simulation-based education for residents. Taking into account the cost of implementing the program, said Okuda, they found that the net annual savings were greater than $700,000 (Cohen et al., 2010). This is a great example of how SOTL can be used to determine the best way to do training and study the impact of training on outcomes and costs, he said. Instead of "training for the sake of training," SOTL can ensure that resources are invested in programs that lead to outcomes with a substantial ROI. Jeffries shared another example in which the value of a simulation-based training program was demonstrated. A major multi-site study of 10 nursing schools looked at whether simulations could replace hands-on clinical time without negative consequences for the nursing students or their patients (Hayden et al., 2014). The study looked at board exam pass rates and clinical practice outcomes after graduation; it had major implications for nursing education both in the United States and globally, said Jeffries. Using evidence from the study, regulators have created guidelines and standards for their states. While every state has different rules, the study suggested that simulations could replace up to 50 percent of clinical time. This example, said Jeffries, demonstrates that SOTL can have a substantial impact on not just HPE but also policies and practices that affect patient outcomes. Through these efforts, she noted, health professionals are getting evidence on best teaching practices and better understanding of how policies and practices impact outcomes.

Challenges of Implementation

Given the potential value of SOTL, Amiel asked panelists about the challenges in implementing it in health professions schools and programs. Jeffries said that leaders—such as school deans—play an important role in propelling SOTL forward and adopting teaching approaches that are grounded in evidence. A leader can question long-standing approaches and suggest new approaches and can facilitate and encourage the key studies to gather evidence on these approaches. Some faculty may resist because they have "always done it this way," said Jeffries, which presents an opportunity to conduct a direct comparison between existing approaches and new ones. She noted that the decision about how to teach a subject should not be based on someone's opinion about what works but rather on evidence from a variety of sources. Sources of data for comparing educational approaches could include student evaluations, patient outcomes in practice, and metrics of successful transition from learning to practice. Success in academia can be thought of as a three-legged stool, said Jeffries; academics have a duty to excel in teaching, research/scholarship, and service. Promotion and tenure criteria are often overly focused on research/scholarship, and faculty often seek funding for research on health and medicine rather than education topics. SOTL can shift these priorities by bringing together the "legs" of research/scholarship and quality teaching. There may be opportunities, said Jeffries, to offer innovative funding to faculty to pursue research on teaching and learning to propel the field of SOTL forward. Okuda added that another challenge in HPE is that faculty and leaders sometimes assume that new technology-based teaching tools are effective and implement them without evidence that they are as good as or better than existing methods. The COVID-19 pandemic accelerated their development and implementation, he said. Okuda said that he is conducting research that compares traditional forms of teaching to tools that use extended reality, to assess whether these tools should be adopted. For example, one study compared the effectiveness of using a two-dimensional screen versus virtual reality. Virtual reality is far more expensive, he said, so there is no reason to use it if it is not more effective. He urged stakeholders to consider the marginal benefit of new technologies rather than adopting something "just because it's cool."

SOTL and Interprofessional Collaboration

Amiel asked panelists to comment on the value of SOTL in the area of IPE and collaborative practice. Okuda said that "everybody knows" the importance of IPE that brings faculty and students together across professions to accelerate and improve learning. However, it can be challenging to measure the actual value of these approaches and particularly to see whether and how IPE impacts practice. This gap can be filled by further research in this area and adequate funding to support it, he said. Nunez agreed about the gap in high-quality educational research on IPE and noted that one of the ways to get this evidence is through SOTL. It is a fundamental component of the relationship between IPE, collaborative practice, and improved health outcomes, said Nunez. It has been demonstrated that high-functioning collaborative teams have a positive impact on health outcomes (Weaver et al., 2014; Wu et al., 2018), she said, and SOTL can be a means of determining and informing the best practices to prepare a collaborative practice-ready workforce. Educational research on this topic could determine the relationship between teaching practices and learning, learning and practice, and practice and outcomes.

SOTL and Faculty

While SOTL has shown value in improving education and ultimately patient outcomes, said Amiel, faculty and institutions may need additional incentives to adopt it. He asked panelists to comment on how its value could be acknowledged and enhanced in the academic world, such as in promotion and tenure. Nunez said that higher education administrators, faculty, and journal editors could better understand the different types of scholarship and their importance to the field as a whole. As Boyer proposed in 1990, scholarship is a broad term that encompasses the individual scholarships of discovery, integration, application, and teaching and learning. Leaders could acknowledge how these categories are interrelated and integrated, said Nunez. SOTL has a "huge value proposition" that is not often recognized, she said, and contributes to improving HPE and ultimately health practice and outcomes. If that connection were better understood, SOTL might be valued more. To make SOTL and its value better understood, said Nunez, it would be helpful to have a common language across disciplines. Working collectively to advocate for SOTL will "raise all boats" and make it easier to gain the support, professional development, mentoring, funding, and recognition that could propel SOTL forward. Huang added that teaching is often framed as a service and an expectation in the academic world. Promotion and tenure decisions are based on other factors, such as impact and reputation, while teaching is sometimes forgotten. Promotion and tenure committees could be better informed that success in SOTL does not necessarily result in high-impact journal publications or a high number of citations, said Huang, but instead is seen as dissemination and adoption in other institutions.

Aside from promotion and tenure, Amiel asked Huang how institutions can support faculty development for those engaged in SOTL. All institutions have faculty development, said Huang, but they do not necessarily have author or scholar development. Scholarship is a critical part of faculty development in elevating one's impact and reputation. Devoting more resources toward educational scholarship could, for example, be used to set up writing accountability groups. Huang acknowledged that professional development is already "resource crunched" but said that an emphasis on SOTL can help recruit and retain talented educators and highlight innovative work happening in HPE institutions.

CLOSING REMARKS

The four categories of scholarship in Boyer's framework—discovery, integration, application, and teaching and learning—are equally important and often interrelated, said Huang. The scholarship of discovery leads to new knowledge, which is expanded through the scholarships of integration and application taught through SOTL. SOTL includes forms of educational research, which is scholarship of discovery, and includes educational innovations in integration and application. Amiel added that traditional research has changed a lot over the past several decades, with a new focus on areas such as implementation science and community-based participatory research. SOTL offers an opportunity to think about how traditional research and education practices have come to be and can be improved, he said.

Jeffries noted that an essential part of a health professional's work is the ability to problem solve and think critically; clinical care has constant uncertainty, and every patient and scenario is different. Critical thinking is a desired outcome in every HPE discipline, and SOTL presents an opportunity to determine and implement best practices for cultivating this skill. For example, said Jeffries, simulations can help students practice using their critical thinking skills in a clinical scenario and then immediately debrief with a facilitator to discuss errors or questions. She noted that finding the time for this in traditional clinical education can be challenging. SOTL is an approach for discovering and adopting such educational innovations to improve HPE and patient outcomes.

Kim Dunleavy, forum representative for the American Council of Academic Physical Therapy and planning committee cochair, closed the preworkshop session by reflecting on some of the main points of panelists and looking ahead to the workshop. One observation she noted is that SOTL can drive innovations to respond to challenges in the workplace, new technologies, and health crises. To leverage the potential of SOTL, Dunleavy commented that it would be necessary to support and train faculty, disseminate information, and collaborate across disciplines. Furthermore, Dunleavy noted that panelists had discussed the importance of showing the ROI of SOTL and the desire for academic leaders to understand and acknowledge its value. Dunleavy expressed that the workshop would highlight and delve more deeply into what it would take for wide adoption of HPE SOTL across all the health professions.

3

Workshop

Zohray Talib, California University of Science and Medicine and forum cochair, opened the workshop and introduced Jeffries and Dunleavy as cochairs of the planning committee. Jeffries shared the framework for the workshop series (see Figure 1-1). She explained that the workshop would focus on two aspects of health professional education (HPE) Scholarship of Teaching and Learning (SOTL): one, its importance and value, and two, innovating and enhancing it. The objective, said Jeffries, is "to shed light on the critical value of SOTL for health professions' workforce development and patient outcomes by exploring what is unique about SOTL in the health professions from education to practice."

Dunleavy gave a brief overview of the history of SOTL that built on the description offered by Jensen in the preworkshop session (see Figure 3-1). In 1990, Boyer advocated for a broader meaning of scholarship, encompassing the scholarships of discovery, integration, application, and teaching and learning. In 1998, Glassick and colleagues highlighted the importance of standards across all types of scholarship, said Dunleavy. His proposed standards included clear goals, adequate preparation, appropriate methods, significant results, effective presentation, and reflective critique. Shortly after Glassick and colleagues's contributions, Hutchings and Shulman (1999) argued that SOTL needs to drive systematic improvements. That is, explained Dunleavy, scholarship about teaching and learning would be publicly available, peer reviewed and critiqued, and reproduced and built on by other scholars.

Boyer's model of scholarship has overlap and interdependence among each of the categories, said Dunleavy. Scholarship of discovery asks the

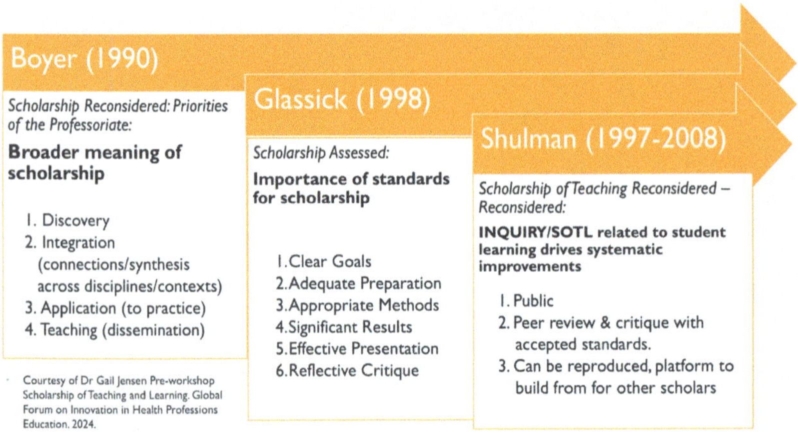

FIGURE 3-1 Historical journey of SOTL.
NOTE: SOTL = Scholarship of Teaching and Learning.
SOURCES: Presented by Kim Dunleavy on October 23, 2024. Courtesy of Gail Jensen.

question, "What is yet to be known?" This area could include HPE research that addresses core problems, basic principles, or foundational questions. The scholarship of integration asks, "What do the findings mean?" This is where collaboration across fields provides context and bridges differences to synthesize knowledge. The scholarship of application looks to answer the question, "How can knowledge be applied to solve problems for individuals and institutions?" This area explores problems in their specific context and drives innovation that is grounded in theory and evidence. Finally, said Dunleavy, SOTL asks, "How can knowledge be accessed and applied? What teaching methods are the most effective for different levels of learners?" Scholars might study teaching, mentoring, development of learners, instructional methods and technology, implementation, and curriculum. While each type of scholarship is different, said Dunleavy, they are interrelated and can be seen as an integrated whole under the broad banner of scholarship.

FOCUS GROUP EFFORTS

Nunez shared details with workshop participants about the focus groups[1] that the planning committee conducted in August. The purpose, she

[1] See meeting materials for a full description of the focus groups: https://www.nationalacademies.org/event/42801_10–2024_scholarship-of-teaching-and-learning-in-health-professions-education-a-workshop#sectionEventMaterials (accessed on February 16, 2025).

said, was to inform the planning with a better understanding of individual and institutional perceptions, beliefs, and attitudes about SOTL and how it is applied in HPE. The four groups had 25 participants from 11 different health professions. They held a variety of roles in HPE, including educators from different graduate-level HPE programs, clinical educators, academic leaders, administrators, and journal editors. Each group was composed of participants with different levels of experience in SOTL, said Nunez; 16 percent self-identified as participants (e.g., read literature, attend workshops), 36 percent as contributors (e.g., publish articles, offer faculty development), and 48 percent as subject-matter experts (e.g., publish extensively, serve as mentor or journal editor). Before the focus groups, the planning committee had identified specific areas of interest within SOTL; the interests of the participants collectively reflected all of these areas:

- Education within one's health profession,
- Clinical care or clinical learning,
- Simulation,
- Interprofessional education or practice,
- Faculty development, and
- Promotion and tenure.

The focus group discussions centered on four topic areas, said Nunez: motivation for engaging in SOTL, institutional support for it, professional association support for it, and building value and innovation for it in HPE. Nunez gave an overview of the questions and responses in each area. Participants were asked what motivates them to engage in HPE SOTL; Nunez identified several themes. Participants explained that they were motivated by the ability of SOTL to demonstrate value and quality, the desire to develop high-quality faculty and support promotion and tenure, the potential for evidence-based education practices to improve outcomes, and the hopes of bridging academic education to clinical care outcomes, said Nunez. When asked whether and how their institutions support SOTL, answers fell into two camps. Those who reported support said that the institution's mission and actions were aligned with SOTL; it was integrated and encouraged as part of merit, promotion, and awards; and faculty were given explicit support and development opportunities. Those who said that their institutions were not supportive reported a mismatch of espoused institutional values and SOTL support (e.g., only research was valued); no evaluation systems for SOTL or consideration of it in promotion and tenure; and limited support, funding, and development opportunities.

Next, participants discussed whether their professional associations facilitate the advancement of SOTL by, for example, offering professional development opportunities. Like academic institutions, said Nunez, support varied. She noted that some associations were less likely to actively support SOTL due to their particular mission; for example, an association focused on developing researchers may not focus on it. According to Nunez, participants identified several approaches to support. These included providing resources, such as materials and support for academic and clinical educators (e.g., dissemination grants, accreditation standards); offering faculty and professional development opportunities (e.g., educational workshops, peer mentoring); and giving recognition (e.g., awards for SOTL achievements and excellence in teaching).

The fourth topic area for discussion focused on what could build value or innovation for HPE SOTL among different stakeholders. Nunez described the participants' suggestions:

- Using Boyer's scholarship model in academia;
- Offering an SOTL track in HPE programs;
- Developing common SOTL language/terminology;
- Enhancing the rigor, designs, and quality of SOTL research;
- Measuring the impact/outcomes of SOTL;
- Improving institutional and structural support, including funding, grants, promotion and tenure guidelines, and merit awards; and
- Developing communities of practice around SOTL, including interdisciplinary collaboration, mentorship, and networks.

During the focus group discussions, said Nunez, several participants mentioned initiatives that could be implemented collectively across HPE to advance SOTL. These are listed in Figure 3-2 and include joint advocacy for HPE SOTL, development of professional identities that combine education and research or education and clinical care, publication of SOTL-related

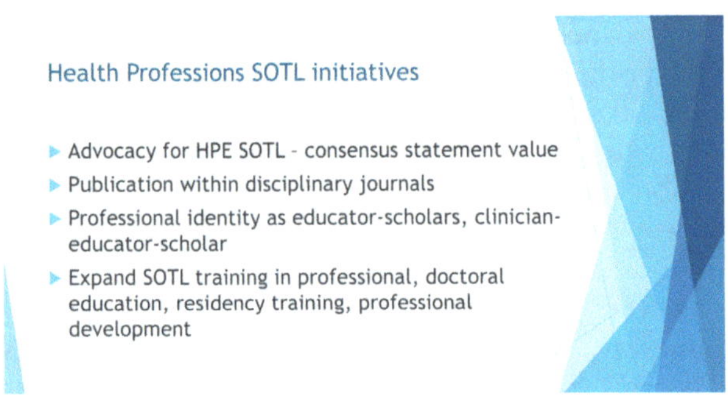

FIGURE 3-2 Potential initiatives to advance SOTL.
NOTE: HPE = health professional education; SOTL = Scholarship of Teaching and Learning.
SOURCE: Presented by Loretta Nunez on October 23, 2024.

material in disciplinary journals, and training in SOTL at every level (e.g., doctoral education, residency training, professional development).

THE VALUE OF SOTL IN HPE

All HPE programs share the goal of preparing learners for professional practice, said Jensen. This shared foundation of education, training, and identity development is sometimes called "knowing, doing, and being" or the "habits of head and hand and heart." SOTL can positively impact each aspect across professions, she said. Improving HPE through SOTL can have far-reaching consequences, starting with improved student learning and leading to improved patient outcomes and community health. Jensen introduced H. Carrie Chen, Kaiser Permanente Bernard J. Tyson School of Medicine, to talk about the value of HPE SOTL.

"Why is HPE SOTL important?" asked Chen. Ultimately, the aim is to improve the health of the public, she said, adding that SOTL supports the best practices in education to train learners to better care for patients by delivering safe, effective, and efficient care. Just as clinical care is ideally grounded in evidence-based practices, said Chen, teaching and educational practices should be too. There are studies that look at the impact of teaching and learning on patient outcomes, and some have identified "educationally sensitive patient outcomes." However, said Chen, there are challenges involved in measuring the impact of teaching; education and health care are complex, so it can be difficult to demonstrate direct cause and effect. Paraphrasing Glenn Regher at the Center for Health Education Scholarship in Canada, Chen said, "HPE research is not rocket science—it is harder." This is due to several factors, she added. Trainees have agency, and educational programs are not able to control their behaviors. For example, a program might inadvertently leave out a significant part of the curriculum but find that learners did just as well on the national exam because they taught themselves the material. In addition, patients also have agency and their own choices and behaviors. Furthermore, patient care is often provided in teams, and it can be difficult to tease out the contributions of individuals. Finally, said Chen, context matters; the impact of educational and patient care interventions may vary significantly based on their implementation contexts.

There are also challenges related to the standards of evidence in SOTL. The health professions are rooted in biomedical science, said Chen, which views evidence from randomized controlled trials (RCTs) as the gold standard. RCTs are generally not feasible in the educational setting and may not be necessary. Chen shared a "funny but real" review article that looked at the use of parachutes and major trauma due to gravity (Smith and Pell, 2003). The review concluded that parachutes should not be used, due to a lack of evidence from RCTs; alternatively, it suggested that in certain circumstances, other types of evidence could be sufficient. By their nature, innovations do not have strong evidence of efficacy, said Chen. It is important for innovations to be theory informed and have efforts to generate evidence substantiating the value of an innovation. However, Chen said that if "we restrict ourselves to practices for which we already have evidence, then we would never innovate." She suggested that due to these challenges, SOTL would most effectively be viewed through not the biomedical model but the models of implementation science and social science. These models value using observations to inform practice, sharing observations and lessons learned, replicating others' work, and teaching others best practices. This process builds on discovery to increase understanding and touches on each of the categories of scholarship defined by Boyer (discovery, integration, application, and teaching and learning). Chen argued for value in shifting the perspective on what would make SOTL worth doing; instead of using strong evidence as the bar, set the bar at improving practice. That can be broadly defined to include improving educational practices, health professions training, and health care itself. Within this frame, she said, different types of SOTL and avenues for dissemination can be recognized and valued. For example, the best dissemination and impact of a practice or innovation may be seen through not a traditional article but an avenue such as a blog.

Given demonstrated value of SOTL, said Chen, how can this value be conveyed (see Figure 3-3)? The first step, she said, is to make SOTL visible and understandable for all of those involved. Chen said that chairs and promotion committees often do not feel capable of evaluating teaching

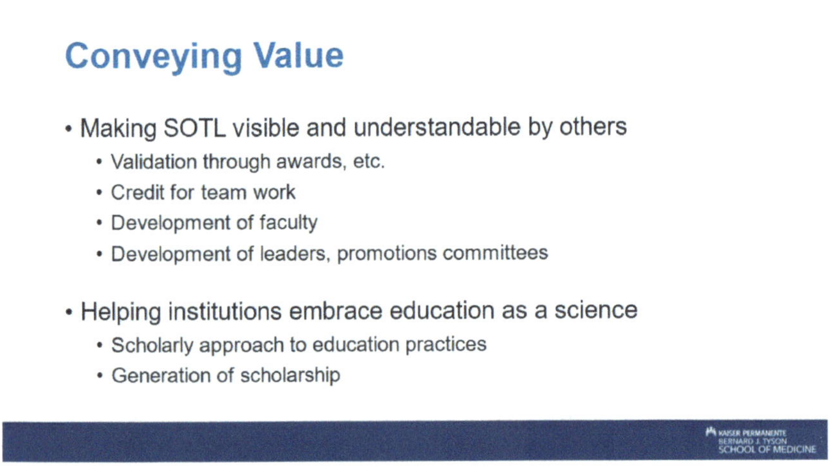

FIGURE 3-3 Conveying the value of SOTL.
NOTE: SOTL = Scholarship of Teaching and Learning.
SOURCE: Presented by H. Carrie Chen on October 23, 2024.

or SOTL excellence. There are ways to make their jobs easier, such as giving awards for teaching or demonstrating the reach of a teaching practice. These approaches provide validation of quality work from education experts. In addition, it is important to teach faculty how to make their own SOTL visible, said Chen. For example, they could track the number of times that a colleague has replicated their teaching practice or approach and include this information on promotions materials. "We need to help institutions embrace education as a science," said Chen, and value the scholarly approach to education practices and the generation of scholarship. An institution that values SOTL would provide faculty development in SOTL, hold faculty accountable for using evidence-informed practices, and provide the resources that faculty would need to generate scholarship related to teaching.

Discussion

Jensen moderated a discussion session and invited workshop participants to ask questions. Jensen began by asking Chen about her personal journey with SOTL and what motivated her to start working to improve educational practices. Chen replied that she first became interested in education from the perspective of a learner. She wondered about ways to make learning more fun, interactive, engaging, and effective and if the policies that shape the learning environment could be modified to account for the fact that learners are human beings with individual desires and needs. Chen said that as she became a faculty member, she learned about a whole field of education with evidence to inform teaching practices. Role models and mentors helped Chen explore this evidence, develop innovations, and learn from mistakes. This process, she said, helped her realize the importance of sharing and helping other educators.

Noting that the participants represented leaders from across the health professions, Jensen asked Chen what she thought was the most urgent action item all leaders can embrace, regardless of their discipline. Chen responded that reflecting on the learning environment and examining how structures and resources within it can impact learning are things that all leaders can promote. Chen then added that faculty sometimes "point the finger at learners" when they are not doing what faculty expect; instead, faculty would need to consider how the environment is shaping learner behavior. Furthermore, institutional leaders can consider how faculty support and promotion and tenure processes can either facilitate or hinder faculty engagement in SOTL. Leaders who want faculty to engage in SOTL would have to provide faculty development opportunities, time and space for scholarship, and mentorship, said Chen. Some faculty who are involved in SOTL report feeling "very alone in their spaces."

Generating Evidence on Outcomes

Tony Breitbach, who represents the National Academies of Practice (NAP) on the forum, recalled Chen's comment about the challenge of measuring team-based outcomes. He mentioned that St. Louis University, where he is employed, hires students who are seeking a Ph.D. in industrial organizational psychology as research assistants. It brings a different frame to HPE; it examines the motivations of groups of people and the contextual factors that contribute to how teams work. This lens, said Breitbach, can help address the challenge of studying team-based activities in HPE. Chen agreed and said that this lens could also be used when looking at how faculty work together and support each other in SOTL. Chen noted that HPE has a view of teamwork that is not always beneficial. Scholars in HPE often follow the model of giving credit only to the first, second, and last author on a publication and also worry about diluting authorship credit by including too many authors. This perspective leads to practices that are not as inclusive as they should be, she said. It is important to think about how to value all the people involved and to do more team science in HPE.

Another participant noted that funders are often focused on patient outcomes and asked Chen for suggestions on how evidence could be generated to connect SOTL to these outcomes. Chen replied that evidence about the impact on patient outcomes is the "holy grail" but that it is very challenging to get these data. There are many intervening factors, and it can be difficult to connect all the dots between an educational practice and an eventual patient outcome. Chen said that it is important to look for evidence that impacts each "dot." For example, researchers could look to demonstrate that a teaching strategy improves trainee learning and then that trainee learning is linked to certain patient care approaches or patient outcomes. It may be necessary to take evidence generation one step at a time rather than aiming for the "holy grail," said Chen.

Chen added that education and training can have outcomes that are both intended and unintended and both positive and negative. It is important to look at all of these outcomes and recognize the complexity of these relationships. Furthermore, impacts may be related to the explicit and the "hidden" curriculum. Chen shared an example from her own work. A problem-based learning course put students into small groups for part of the curriculum. Chen looked at change in student behaviors and found significant variation among groups. She noted that this could have been due to the facilitators but also could have been due to level setting among the group participants. This demonstrates that the impact of a learning experience is based on not just the high-level curriculum but also how it is implemented, delivered, and experienced. Translated into the patient care environment, this phenomenon means that outcomes can be influenced by both the individual team members providing care and the team dynamics.

Funding

Lyuba Konopasek, the forum representative of Intealth, asked Chen for her thoughts on the lack of funding for SOTL and how this situation might be improved. Chen responded that a lot of work in SOTL is being funded in certain places worldwide—specifically, the Netherlands and Canada. Looking at how these countries began to prioritize SOTL may help forge a path for better funding in the United States, she said. Chen noted that the largest funder of educational research is the Department of Defense; she asked the audience how health professional educators might begin to think differently about education and inspire society to value it in a different way. As a participant, Talib reflected on a global investment in HPE. The Medical Education Partnership Initiative was cofunded by various federal agencies, and it invested $150 million in HPE in sub-Saharan Africa. This money, she said, transformed the landscape of HPE in that area. Unfortunately, the grant expired after 5 years, but the systems investments in faculty lasted beyond that period. Talib noted that investments in the classroom tended to cycle out quickly, but those in faculty had more sustainable benefits. One of the universities involved wrote a paper that laid out a logic model for how HPE contributes to the health system (Mokone et al., 2014). Studying how educational systems contribute to health outcomes could support investing in HPE SOTL, said Talib. "If we can make the case, then maybe the funding will follow," she said. Chen added that conversations are happening about how the Kaiser Permanente Bernard J. Tyson School of Medicine invests money and whether some approaches have a "bigger bang for the buck." A data gap exists in this area, said Chen, but it is important to look at the impact and sustainability of investments at all levels.

Participants also commented briefly on a number of other topics for consideration:

- Lessons from SOTL in other fields, particularly those that work in teams;
- A desire to change the culture of HPE academia so faculty are valued who do nontraditional work, including SOTL and team-based education; and
- The importance of teaching learners to provide compassionate care, given evidence that patients' outcomes are affected by their feeling that their health care providers care for them.

Chen concluded by noting that it can be "overwhelming" to think about how to change the culture in HPE to better understand and value

SOTL and said that it is critical to have these kinds of interdisciplinary crowdsourcing sessions to discuss how to move forward.

BREAKOUT ACTIVITY: THE VALUE OF HPE SOTL

Participants broke out into groups. In-person attendees remained in the room for table discussions; virtual attendees watched the evidence-based videos from SOTL subject-matter experts[2] and responded to questions in an online survey. Questions for discussion included the following:

- What are the elements related to HPE SOTL that are critical and make it unique?
- How do we connect HPE SOTL from classroom to clinic?
- What strategies would you use to bring HPE SOTL value back to your profession and/or across health professions?
- How do you advocate for implementation of SOTL innovation?
- How do we promote HPE SOTL as part of professional identities?
- How do you expand HPE SOTL training within and across health professions?

After the breakout groups concluded their work, participants reconvened for the second day of the workshop. Huang offered reflections on the first day, including takeaways from breakout group discussions. Huang said that participants mentioned a number of inherent tensions in the area of SOTL, such as seeking rigorous evidence while making research feasible, reconciling the desire to demonstrate a relationship between SOTL and patient outcomes and the reality that using process measures is more achievable. Furthermore, participants discussed holding a preference for a high burden of evidence while navigating the fast-moving pace of innovation. Huang noted that it may be advantageous to shift away from a biomedical model of evidence and toward methodologies such as implementation science.

To move SOTL forward in HPE, Huang said, participants discussed building bridges to stakeholders' unique needs and roles. She added that faculty need education on the importance of SOTL; promotion and tenure committees need to understand its value; and educational leaders need to offer institutional support for it. To extend progress further, investments should be focused on the professional identity formation of HPE faculty as scholars, while continuing to emphasize the critical role of teaching in

[2] See video resources at https://www.nationalacademies.org/event/42801_10-2024_scholarship-of-teaching-and-learning-in-health-professions-education-a-workshop#sl-three-columns-e5a903f9-5ee2-4a45-89d1-d9476c8c1783.

improving the health of communities. Rather than each discipline within HPE taking on this work individually, Huang said that bridge-building would best be accomplished through interprofessional collaboration. For example, professional societies could coordinate efforts, a national registry of learner outcomes could be developed, or a national institute for HPE SOTL could be established.

This workshop has been a "celebration of how wonderful and not in the box HPE SOTL is," said Huang. SOTL is by its nature interdisciplinary and interprofessional, and it requires a team science approach. It involves both quantitative and qualitative methods and invites nonmedical disciplines to participate. SOTL illustrates the "messiness" of research in HPE, said Huang; it is not simple or straightforward to connect the dots between education and patient outcomes, but it is necessary and valuable work. Dunleavy added her thoughts on the workshop discussions thus far and said that SOTL draws attention to areas of health care practice that are critical but often overlooked. Skills like communication, empathy, and teamwork are incredibly valuable in health care, said Dunleavy, but how do health professional educators teach these skills, and how do we measure them? Moving forward, she added, educators and HPE SOTL researchers could identify the skills that make a difference in patient outcomes and determine how these are best taught to learners. SOTL is a way to connect best practices in HPE programs with the ultimate goal of improving patient outcomes, Dunleavy concluded.

INNOVATIONS IN HPE SOTL

Dunleavy introduced Patricia O'Sullivan, University of California, San Francisco, School of Medicine, to discuss innovations and new ideas in HPE SOTL. O'Sullivan shared some opening remarks, after which Dunleavy moderated a question-and-answer session. Innovation starts with how health professional educators approach the research, said O'Sullivan. HPE SOTL is a "team sport," and it requires collaboration between experts in SOTL and faculty members with a passion for HPE. In addition to drawing on multiple disciplines, innovation in SOTL requires conscious and intentional inclusion of a broad range of voices. O'Sullivan noted that people often simply "turn to the people who are around" when embarking on a project rather than reaching out to find the voices that are missing. Innovation in HPE SOTL will not happen without "relentless" networking and moving out of silos, she said. Despite increasing scholarship in the area of HPE SOTL, it is made up of a small group of authors who often cite each other. Connecting to other disciplines, such as education, psychology, and sociology, would open the work to new perspectives, methodologies, and frameworks that could enrich the work. Furthermore, said O'Sullivan, so would reaching out to scholars in other areas of the world who are often not part of the HPE literature. O'Sullivan noted that networking can and ideally would also happen in one's own institution. Faculty who are engaged in SOTL can reach out to other faculty and leaders to educate, support, and motivate. O'Sullivan shared a story about a division chief who was hesitant to give financial support to an educational scholar. The department chair spoke up and said that it was an "issue of equity," pointing out that there would be no question if it were a clinical researcher asking for funds. O'Sullivan encouraged people engaged in SOTL to work with leadership and help others in asking for the support and resources that they need.

Discussion

Dunleavy and workshop participants engaged in discussion with and asked questions of O'Sullivan that fell into four general areas: faculty development, alternative means of dissemination, challenges in SOTL, and measuring learning.

Faculty Development

In Europe, those who teach in higher education are required to have a basic teaching background. While this concept has not been embraced in North America, it is important to think about what health professional educators can learn from their system, said O'Sullivan. One approach for preparing faculty with the right skills, she said, is longitudinal faculty development programs. Evidence supports the effectiveness of these types of longer-term development programs; this is where the development of identity and change can occur. Another thing that is happening, said O'Sullivan, is the growth of masters' and doctoral degrees in HPE. These programs develop valuable scholarship and deep thinking. It is important to have people in the HPE environment who can ask questions such as "What is the theory driving this curriculum?" or "How do we understand and adapt to our learners' motivations in order to improve patient outcomes?" There has been significant progress in education in the workplace in HPE, said O'Sullivan, from simulation to skills practice. Quality faculty development programs can expand on this to prepare learners to provide the best patient care.

Alternative Means of Dissemination

Dunleavy noted that workshop speakers talked about alternative dissemination methods for SOTL, outside of the traditional journal system. She asked O'Sullivan about these and how to ensure appropriate review and rigor. O'Sullivan responded that many options within the journal world are often overlooked. For example, she has a Korean colleague who publishes in a Korean journal to bring her ideas to a community that may have difficulty reading English journals. This would be important to explain to the members of a promotion and tenure committee, she said, because they may not understand the value of the contribution to that community. Other journals with targeted audiences (e.g., military medicine) could be useful for disseminating information about SOTL. There is great potential in alternatives to journals, such as blogs, online curricula, websites, and podcasts, said O'Sullivan, but it will be important to develop guidelines to ensure their quality. The criteria do not need to be the same as those used in journals, and different methodologies could be used to determine "quality." O'Sullivan gave an example of an alternative method: one of her colleagues used a large language model to create a dialogue between two individuals about a journal article they had written. The core information was the same, she said, but it was presented more engagingly.

One major issue for faculty working in SOTL is how to convey their work understandably to a traditional promotion and tenure committee, said O'Sullivan. The standard format of the curriculum vitae is often merely a reflection of what has been done and does not show the impact or the quality of work. O'Sullivan encouraged faculty to explore different ways of describing their work and "seize the opportunity" to highlight the value of SOTL work.

Challenges in SOTL

Efforts have been put toward helping individuals in SOTL, said O'Sullivan, but the field will not move forward without work on the systems level. Students, postgraduates, and faculty engaged in SOTL are all struggling with the same issues, indicating that work within the system—the education to practice continuum—could help build a system that values SOTL. If educational and health care institutions value it, more resources and tools will be available, and people will be able to focus on quality work over quantity. Educators need the space to explore how to best prepare learners for the future rather than trying to produce more articles or make more quantitative impact. The challenges for individuals working in SOTL will not go away until the system itself values it.

Dunleavy said that the focus groups revealed a lack of support for SOTL at many institutions; she asked O'Sullivan for her thoughts on how to move it forward in these circumstances. O'Sullivan replied that one way to get started is to align SOTL with institutional values. Every institution has values and issues that they care about, she said, and showing how HPE SOTL fits into this framework can create a pathway forward.

Jensen asked O'Sullivan about the role that status plays in keeping SOTL from being more broadly adopted; she noted that calling it "education science" can bring more visibility. O'Sullivan said that SOTL could benefit from using more design-based research. Some research on educational strategies is atheoretical and makes a lot of assumptions about why and how teaching interventions work. In contrast, design-based research requires understanding the theoretical principles underlying the intervention and enables rapid prototype testing. The researcher starts by hypothesizing the outcome, based on theoretical principles, and tests the hypothesis. If the outcome is not achieved, the researcher works with the learners to determine what happened and how the intervention could be redesigned. Design-based research, said O'Sullivan, has no need to wait years to see if the learning outcomes were achieved. Instead, there is constant testing and feedback about the process of learning.

Measuring Learning

A participant noted that research on educational innovation is episodic, whereas learning is a process that happens over time. He asked O'Sullivan how research in this area could better assess learning across contexts or over time. She responded by encouraging educators to think long term and strategically about how to gather data on learning outcomes. Some institutions have established longitudinal databases, and some schools in Europe use portfolio systems that enable examining how someone is learning over time. Longitudinal qualitative research has also emerged, said O'Sullivan, which takes a team commitment to slow down and aim for quality over quantity. O'Sullivan relayed a story from her colleague, who asks would-be researchers to describe not just their first proposed study but also follow-up studies that would build on its findings. "If the researcher can't lay out three studies in a row," she said, "they shouldn't do the first." While this might be a "stressful" way to think about research, O'Sullivan said that a coordinated and long-term study plan will contribute much more to the collective understanding of SOTL than a series of one-off research projects.

BREAKOUT ACTIVITY: FACILITATING CHANGE USING HPE SOTL

Participants divided into groups for an activity. Virtual attendees responded to the question "How would you develop a program of the future to facilitate change using HPE SOTL?" They were asked to name key elements in 300 characters or less. In-person attendees remained in the room for an activity led by forum members. The groups were asked to consider

1. the present state of HPE and workforce; and
2. the potential for new tools, technology, and methods to change

how students and health professionals learn and how clinical and classroom educators teach.

Each group was tasked with illustrating an innovative HPE SOTL design that would produce a meaningful result(s) for a target population and describing the relevant stakeholders. Groups created visuals to summarize their responses. Afterward, representatives from each group gave a report on their work, and Okuda moderated a discussion among participants.

Group 1

Nancy Moreno, Baylor College of Medicine, reported that Group 1 had a lively discussion about how to form a community of practice around SOTL in an organization. The figure they created (see Appendix B) uses the model of a house to illustrate their idea. Moreno noted that the house has three stories but emphasized that their model is not hierarchical. Rather, the base of the house is foundational knowledge, and the upper floors contain rooms with thematic areas related to SOTL, including inquiry, dissemination, and implementation, and a person who represents an expert in that area. Moreno explained that in community of practice theory, experts bring novices into the community through shared knowledge, understanding, and skills. The sun shining on the house represents institutional support, and outcomes emerge from the chimney and impact the greater community. Inhabitants of the house, said Moreno, include faculty, new faculty, existing faculty learners, experts who are specialists in SOTL, and other members of the community who share the goal of supporting excellent care for patients.

Group 2

Group 2 chose to focus on novice faculty, said Jeffries, on the clinical or academic side. With the goal of increasing knowledge and use of SOTL to improve scholarly teaching, their model (see Appendix B) illustrated a learning hub that included theories, literature, exemplars, and model applications. Learning is multidirectional, and learners are engaged in the process. Jeffries said that the group was creative in the tools for the learning hub and included innovations such as artificial intelligence (AI) and holograms. The content would not be solely focused on SOTL but would also include HPE topics. For example, a faculty member might give a cardiovascular lecture, after which another scholar could offer peer review and a debriefing dialogue. Other interventions could include mentorship and coaching, simulations, and paired novice–expert teaching. Learner feedback and reflection are essential parts of the model, said Jeffries; faculty would not simply "assume that they did a good job." A circle around the figure represents several foundational aspects of the model, including evidence-based practices, grounding in educational theories, and a sense of psychological safety. Jeffries explained that the latter is essential because it can be difficult for novice faculty members to be critiqued.

Group 3

Group 3 had members from many different professions, said Katie Eliot, forum representative from the Academy of Nutrition and Dietetics, and quickly identified the shared challenge of learner engagement. In their drawing (see Appendix B), the disengaged learner is sleeping, "shopping for shoes online," and bored with the status quo of teaching. The group focused on how to use SOTL to transform that person into an engaged learner, such as by changing the learning environment. Jensen added that their model would be an interdependent process for which both faculty and learners would need to be prepared. It is critical, she said, to set the stage and help all stakeholders understand the expectations and the structure of co-participation and build trust and respect. Faculty would play a critical role by advising, mentoring, and coaching.

Group 4

Andrea Pfeifle, one of the forum representatives for NAP, said that Group 4 focused on learners in HPE working in underserved areas. The intended outcome was ensuring that they have the same resources and access to continuing professional development as other learners. Breitbach, also from NAP, explained the group's figure (see Appendix B). It illustrates an academic–practice partnership that offers a continuing professional development program, working with accreditors and informed by accreditation standards. The program would use intervention tools such as AI, electronic health records, and IPE. With the goal of enhancing the skills of health care workers in underserved areas, he said, measured outcomes would include improved teamwork, quality of care, patient–provider experience, and equity across trainees in different areas. Scholars with expertise in each area would pull system data from accreditors to be fed back into the system for continuous quality improvement, sent to the accreditor, and perhaps used to update the standards. Hopefully, said Breitbach, the system would demonstrate success in improving quality, justifying the use of resources for the interventions, tools, and measurements.

Virtual Group

Huang aggregated the comments from hundreds of virtual attendees who participated in the breakout group activity through Slido, the question-and-answer platform. The group's input focused on "out-of-the-box" thinking regarding the future of HPE (see Appendix B). Participants mentioned using innovations such as AI and other technologies to provide real-time feedback and continuous assessment. One commented that AI could be used for teaching basic knowledge and skills, while humans focus on human aspects of HPE that AI cannot teach. Participants also brought up creating linkages and resources among different disciplines within HPE, said Huang. For example, a white paper on SOTL for promotion and tenure committees could be created and shared with faculty and leaders across many different health professions programs. Finally, participants talked about new paradigms within HPE, such as using technology for more individualized learning journeys. Huang concluded that these innovations would undoubtedly cost money, so a dollar sign representing funding was drawn on their model.

Discussion

Okuda invited workshop participants to join him in a debrief of the group activity, sharing their thoughts and feelings about the process of collaborating. Peter Cahn, Massachusetts General Hospital Institute of Health Professions, replied that his group started by discussing the importance of innovation but fell back on the familiar tools they knew when building the model. At the end, they wanted to add innovation back in but could only think of technological innovations. Cahn wondered about other options and how HPE SOTL could use these rather than "defaulting to what's shiny and beeping." Eliot commented that although her group was made up of many different disciplines within HPE, they all had a common challenge: learner engagement. While no one had the "right" answer, a lot of energy and synergy were created by thinking together about that "thing that we all struggle with." Finding these commonalities allows stakeholders from different areas to work together and build on each other's ideas. Sue Forneris, from the University of Minnesota, added that stakeholders across disciplines could benefit from a shared mental model of HPE SOTL and being able to convey it to others. Awareness and articulation of a shared model allows people to work together across disciplines and create synergy, she said.

REFLECTIONS

In the final session, Dunleavy moderated a discussion among planning committee members and participants. She asked people to reflect on the discussions and offer their thoughts and comments. Dunleavy said that one thing that stood out for her was the idea of a white paper for promotion and tenure committees. This is a "very tangible, very much-needed resource." Integrating the knowledge and value of SOTL into these committees can take a long time and a lot of people; a shared white paper would streamline some of this. Cahn agreed, noting the work to do in this area and that systems that recognize faculty tend to hew to the metrics of the biomedical model, valuing quantitative work and randomized trials. Cahn encouraged the group to push back against this idea of "only one form of valuable knowledge" and consider creating space for the value of SOTL to flourish. Breitbach added that he supports the idea of the shared white paper and that having a resource on SOTL would be tremendously valuable for both the committees and the faculty submitting their work. These committees want to do the work and understand this area better, he said, and faculty could use the white paper as a "road map" to make their case for tenure. Another speaker noted an opportunity to create a curriculum blueprint for faculty development in SOTL that could be shared across professions. On this topic, a virtual participant commented through Slido that the University of Washington created an appointment, promotion, and tenure toolkit to help those who identify as team scientists succeed in a system that is based on evaluating individual work. Other participants noted that professional associations have done some work in the area of faculty development in SOTL and credentialing for educators. For example, Katie Sheats reported that the New Academy of Veterinary Educators proposed a model for a credential program that would include portfolio work, experiential-based training, and an examination. The National League for Nursing, said another attendee, has a competency-based certificate process for nurse educators. Dunleavy noted that the education and training time for some health professions has increased in recent years, and this time commitment may prevent some from pursuing additional education-specific certification. Jody Frost, who cochairs the forum, said that one approach for faculty development in SOTL is faculty residency programs; people interested in becoming educators can learn about what is involved with teaching and gain competencies in SOTL.

SOTL is an important topic, said Jeffries, as evidenced by the hundreds of people who signed up to attend the workshop virtually. There is a lot to do, but the ideas, strategies, and synergy at the workshop will be of great value going forward, she said. Frost agreed that SOTL is a topic that is "making a lot of noise," and she suggested capturing the momentum of the meeting by creating a community of practice with all those who attended in person and virtually. Talib followed up with her observation that the energy at this workshop is in part due to the fact that many participants are "hustling in silos in our own institutions." Many people are used to working on SOTL on their own, and they find strength in the group when they all come together. She said that this is another reason why a global space, or community of practice, would be beneficial. Okuda added that it is also important to engage people outside of the SOTL world in order to move things forward. Peer-reviewed journals are one approach for disseminating information, but alternative means, such as blogs, can be incredibly powerful as well. Mark Merrick, forum representative for the Athletic Training Strategic Alliance, agreed that it is important to engage others in the conversation around SOTL. This workshop is "an opportunity to preach to the choir," but structural barriers must be overcome, related to the culture, mission, and practices of institutions, so educational efforts can align SOTL with institutional values. Until institutions value SOTL as a critical piece of their work and put it into practice, "we'll continue to be a choir rather than an entire group singing together." Another participant agreed and said that when something is ingrained in an institutional mission statement, it becomes part of strategic planning and resource allocation. To make this shift, it would be necessary to make the case for why SOTL matters; those involved in SOTL could demonstrate how investments in it translate into meaningful outcomes. These outcomes might be related to community health, education, patient safety, or other areas, she said, but it is essential to connect the dots in order to drive change.

Dunleavy closed the workshop by thanking the planning committee, speakers, staff, and participants. She emphasized the importance of being a "visionary" for HPE and coming together with like-minded people to collaborate and think together about how to move ahead.

4

Closing Session

Frost welcomed participants to the closing session of the workshop series, thanking sponsors, partners, and the planning committee. She introduced Dunleavy to provide an overview of the activities. Dunleavy reminded participants of the objective of the series, which was "to bring forth the critical value of the Scholarship of Teaching and Learning (SOTL) for health professions' workforce development and patient outcomes by exploring what is unique about SOTL in the health professions from education to practice." The preworkshop session, focus groups, and author videos informed the main workshop, which was attended by over 900 people both in person and virtually. The interest in the workshop illustrates the importance of the topic of SOTL across multiple countries and disciplines, said Dunleavy.

In the preworkshop session, speakers and participants discussed the origin and evolution of SOTL, including the historical works of Boyer, Glassick, and Shulman. The first panel covered a number of topics, said Dunleavy, with an emphasis on the scholarship to scholarly teaching continuum, value of faculty development for understanding and implementing SOTL, importance of grounding teaching practices in evidence and theories, and potential to use new methods and technologies for both teaching and dissemination of SOTL. Online participants added to the discussion by suggesting an examination of current trends in education, outcomes, and unintended consequences rather than simply adopting practices because they are the "newest and best." Preworkshop participants were asked to identify critical priorities for building SOTL value and innovation, said Dunleavy, and based on that input, she extracted four areas: professional development, funding, promotion and tenure, and outcome measures. Dunleavy also shared comments from the focus groups held in September. One of the main takeaway messages for her was the importance of elevating SOTL to be accepted as equal to other forms of research. Individual focus group participants also commented on how to structure SOTL to be responsive to changes in health professional education (HPE) and health care, the value of using and developing evidence-based teaching strategies, and the obligation to consider patient outcomes and societal impact along with student learning outcomes.

Huang gave an overview of the presentations and discussions at the main workshop. The ultimate goal of SOTL is to improve the health of patients and society, she said, requiring SOTL to be grounded in strong evidence and to effect practice change (see Figure 4-1). The first speaker, Chen, emphasized the value of shifting away from the biomedical paradigm and embracing the "messiness" of SOTL. Rather than looking for hard clinical outcomes in tightly controlled environments, SOTL uses systematic observation and replication and looks for outcomes that are educationally sensitive. Instead of randomized controlled trials (RCTs) and an emphasis on the rigor of evaluation, said Huang, SOTL leans on social science methods and values innovation and experimentation.

Workshop participants engaged in discussion on augmenting the value of HPE SOTL, said Huang, and covered a number of potential activities in different areas. Expanding and embedding SOTL within HPE may require the following:

- Skill development, including training, mentorship, and formal professional development;
- Advocacy for SOTL and widespread recognition of its value;
- Guidelines for evaluating SOTL work for promotion and tenure;

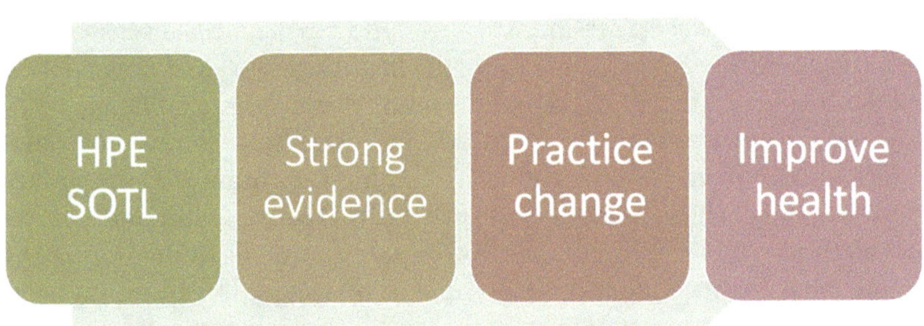

FIGURE 4-1 Value equation for SOTL.
NOTE: HPE = health professional education; SOTL = Scholarship of Teaching and Learning.
SOURCE: Presented by Grace Huang on October 24, 2024.

- Interprofessional and interdisciplinary collaboration and communities of practice; and
- Support structures, funding, infrastructure, and resources.

The second day of the workshop focused on innovation in HPE SOTL, said Huang. After a presentation by O'Sullivan to lay the foundation, participants engaged in a group activity to brainstorm how to leverage innovations in SOTL to facilitate change. Some of the innovations that were identified included the following:

- Leveraging learning hubs to connect across professions and institutions,
- Developing SOTL faculty competencies and accreditation standards,
- Using electronic health record data to measure impact of SOTL,
- Leveraging technology to create individualized learning journeys, and
- Using artificial intelligence for foundational HPE training.

Huang shared a quote from an online participant, which she said summed up the message of the workshop: "Educate academic leaders about the validity and importance of SOTL and give researchers the support needed to do the work." To advance HPE SOTL, said Huang, we must connect HPE with SOTL; this is a simple yet critical proposition for improving health outcomes.

DEFINING MECHANISMS FOR FACULTY DEVELOPMENT

From Jeffries's perspective, one of the major points taken from the workshop series was the importance of faculty development for HPE SOTL. Jeffries reviewed the conversations on this topic and presented her six specific mechanisms that could leverage it. First, institutions could align their missions with the tenets of SOTL. Jeffries suggested that professional and health care organizations and academic institutions could offer grant funding for SOTL projects that are cross disciplinary and collaborative. Second, faculty could participate in team projects that bring together experts in HPE and experts in SOTL to develop initiatives to improve learning and outcomes. Third, faculty could consider longitudinal projects that evaluate new approaches for education in HPE by collecting data throughout the educational journey and post-graduation. Fourth, faculty development could happen via identifying resources and mentors in other departments and across professions. Jeffries noted the shared core competencies among different HPE disciplines, and faculty could work together across professions on SOTL projects related to these. Fifth, faculty could participate in community of practice activities, such as brown bag discussions and group projects, so that interested faculty could share ideas, disseminate innovations, and develop projects. Such a community could be convened virtually, said Jeffries, to allow for participation of a broad group. Finally, faculty could seek out mentors and opportunities in professional associations and in other disciplines. Jeffries suggested that an organization could put out a call for SOTL mentors who could form a group to help others understand SOTL and promote SOTL opportunities.

ELEVATING HPE SOTL

To bring together all the input and discussions from the preworkshop session, focus groups, author videos, workshop, and online comments, Jensen, Huang, Nunez, and Okuda each described ideas that they believed would help elevate and progress HPE SOTL (see Box 4-1).

BOX 4-1
Highlights from the Presentations of Individual Speakers on Progressing Health Professions Education on the Scholarship of Teaching and Learning (HPE SOTL)

1. **Develop strategies that can address and overcome barriers** to systemic change given the complexities of HPE SOTL, including professional silos, lack of widespread understanding, limited resources, and challenges in conducting research. (Jensen)
2. **Craft consensus-based guidelines for promotion and tenure committees** and institutions on how to recognize the quality, value, and scholarly impact. (Jensen)
3. **Make a compelling case for the value of HPE SOTL** within institutions and across professions that can be shared broadly as a unified message to disseminate information on and about it to stakeholders. (Huang)
4. **Create communities of practice and faculty development supports** within institutions and professional associations and across interprofessional and global boundaries. (Nunez)
5. **Identify methods to support alternative forms of high-quality dissemination**, such as blogs, media, and online materials. (Okuda)
6. **Generate strategies to create new funding mechanisms** and infrastructure support for SOTL within and across institutions and professions. (Okuda)
7. **Explore creating a national center on SOTL** or institute on development of teaching and learning and SOTL in HPE. (Nunez)

This list is the rapporteurs' summary of points made by the individual speakers identified, and the statements have not been endorsed or verified by the National Academies of Sciences, Engineering, and Medicine. They are not intended to reflect a consensus among workshop participants.

Strategies to Addressing and Overcoming Barriers

Jensen led off with her ideas for developing strategies to address and overcome barriers to systemic change given the complexities of HPE SOTL, including professional silos, lack of widespread understanding, limited resources, and challenges in conducting research. She noted the lack of knowledge and understanding of SOTL, which impacts decision making and limits coordination and collaboration. This is the place to start in addressing the barriers to systemic change, she said, because knowledge and understanding are foundational steps for addressing any of the other barriers. To begin improving both, HPE professionals would need to work as a team with other stakeholders, including experts from within their own discipline, across the health professions, and in areas such as psychology, sociology, and education. "HPE SOTL is a team sport" and requires interprofessional partnership and collaboration, said Jensen. Those who have expertise in SOTL can partner with those who lack knowledge to share skills, learn, and publish, and communities of practice can cross disciplinary boundaries. Working within professional silos can result in incremental change, Jensen explained, but major systemic change will require all health professions to work together.

Developing Consensus-Based Guidelines for Promotion and Tenure

Her second idea for progressing HPE SOTL would be to develop consensus-based guidelines for promotion and tenure committees and institutions on how to recognize its quality, value, and scholarly impact. Promotion and tenure decisions often focus on quantity rather than quality, said Jensen, and lack an understanding of SOTL and what comprises quality SOTL work. To facilitate a shift, Jensen believes it is imperative to make the case that SOTL prepares future health professionals to provide safer, more effective, and more efficient care to patients. This will require evidence about the relationship between SOTL and patient outcomes, which is unlikely to come from RCTs; borrowing designs and methodologies from social science or implementation science can help generate quality evidence. Finally, said Jensen, shifting the way that promotion and tenure committees think about SOTL will require a culture change. Guidelines and formal rules are not sufficient; faculty and peers have to understand the importance of SOTL and embed it in their work. "The need to train learners and faculty is essential to get strong HPE SOTL promotion and tenure guidelines," said Jensen.

Making a Compelling Case for the Value of HPE SOTL

A third suggestion for progressing HPE SOTL came from Huang, who said to make a compelling case for its value within institutions and across professions that can be shared broadly as a unified message to disseminate information on and about it to stakeholders. Huang reflected on the many attendees at the workshop, representing learners, faculty, institutional leaders, patients, and society, who could benefit from learning about the value of HPE SOTL. While it may seem obvious that SOTL and quality teaching are necessary for quality health care, she said, there are a number of reasons why it is still necessary to argue for it. First, quality health care is taken for granted in some settings and not accessible in others. Second, health care is under threat due to issues such as unaffordability, health disparities, destabilization of the industry, and high operational and labor costs. Third, the stakes for individuals and public health are enormous—and universal. Finally, achieving quality, equitable health care is dependent on high-quality, evidence-based, disseminated training practices for health professionals. Huang emphasized that the message supporting HPE SOTL would ideally be unified across often-siloed HPE disciplines and that collective efforts would lead to a higher impact. Huang identified a number of potential activities that she believed would elevate and progress HPE SOTL:

- Simultaneously published journal editorials on it across health professions,
- A widely disseminated white paper written by a reputable supra-health professions entity,
- An interprofessional or interdisciplinary consensus statement posted on or linked from professional organizations' websites,
- Infographic campaigns in advance of a major proposal or SOTL event, and
- Distribution of a memorable framework that serves as a typology of the SOTL business case for all health professions to refer to.

Creating Communities of Practice and Faculty Development Supports

The fourth idea was presented by Nunez, who suggested creating communities of practice and faculty development supports for HPE SOTL within institutions, within professional associations, and across interprofessional and global boundaries. The former, said Nunez, are typically intimate groups of faculty who meet a few times each semester to talk about teaching and learning best practices and challenges. These groups offer a unique faculty development opportunity by facilitating peer-to-peer learning and knowledge sharing among individuals with common interests or needs. They allow participants to solve problems, compare best practices, and collectively improve their skills by leveraging and sharing their different experiences and perspectives within a supportive environment. Those who could participate and benefit regarding SOTL include faculty, staff, postdocs, and graduate students. One important aspect of a community of practice, said Nunez, is the participation of both subject-matter experts and novices to allow for shared learning opportunities.

Several initiatives could be duplicated, scaled up, or enhanced to create communities of practice for SOTL, added Nunez. These include programs within institutions, such as centers for teaching and learning, and SOTL mentorship programs (e.g., Georgia Southern University SOTL Mentorship Program). Professional associations have initiatives such as special interest groups, communities of practice, and learning communities. For example, said Nunez, the American Speech–Language–Hearing Association (ASHA) Learning Community for Teaching Foundational Communication Sciences was a program that combined a symposium on evidence-based educational practices with a year-long community of practice for participants to discuss and share their experiences implementing what they learned at it. Across professions, associations, and institutions, collaborations such as the Grant Writing and Mentorship in Education Research program exist. This program, Nunez explained, was started by the American Physical Therapy Association and its affiliated education and research organizations. After a few years, it became an interdisciplinary effort by adding the American Occupational Therapy Association and ASHA. At the first interprofessional workshop, 21 participants across the three professions received feedback and mentoring on their proposed research education projects and grants. By expanding or leveraging such initiatives, stakeholders can create communities of practice to build SOTL support, value, and innovation, said Nunez.

Supporting Alternative Forms of High-Quality Dissemination

Okuda described a fifth idea, identifying methods to support alternative forms of high-quality dissemination (e.g., blogs, media, online materials). A lot of great educational researchers are doing great work that often does not get disseminated to the end user who needs it, Okuda noted. In part, this is because SOTL research is disseminated to and consumed by other SOTL researchers rather than the health professions educators who could implement it. In addition, said Okuda, a lot of the information that is widely available in nontraditional forms (e.g., online) is not high quality. "High-quality dissemination requires high-quality work to disseminate," he said. A perennial challenge, however, is achieving this with limited time and money. Okuda suggested that with limited resources, it may be appropriate to use outcome measures that look at incremental steps rather than trying to demonstrate the full impact from education to population health in one article. He noted that interprofessional education (IPE) has faced a similar challenge of being asked to prove that it has an eventual impact on patient care. A collection of high-quality research that demonstrates impact on a smaller scale can eventually add up to showing large-scale impact, said Okuda. These incremental studies, however, may not be able to be published in peer-reviewed journals. Okuda encouraged researchers to use nontraditional means to get their findings out in "bite-size" pieces that are understandable and implementable by educators. It is important to strike a balance between providing information that is sufficient to make the work scholarly and potentially replicable by other educators but not so overwhelming that it is not picked up by busy health professionals and educators. Quality is at the foundation of any valuable research material, whether it is disseminated through a traditional article, or a blog, newsletter, or podcast. "Without quality," said Okuda, "the choice of dissemination routes is really not important."

Creating New Funding Mechanisms and Infrastructure Support

The sixth suggestion was also presented by Okuda and involved generating strategies to create new funding mechanisms and infrastructure support for SOTL within and across institutions and professions. The strategy for creating new funding mechanisms begins with building a business case, said Okuda. Ideally, all the stakeholders for HPE SOTL—faculty, learners, leaders, and patients—would buy into its value. However, it is important for people to see the direct benefits, or the likelihood of engaging is severely diminished, he said. For faculty, the benefits of SOTL might be more effective teaching methods that improve student learning. For students, it could lead to higher grades and improved competency. For leadership and institutions, embracing it could bring about diminished staff turnover, less stress and burnout, and higher student satisfaction. "It is up to each one of us," said Okuda, to identify key players and find ways to demonstrate the value of SOTL for each person. Funding will follow if leaders are convinced of the value and return on investment (ROI). He shared an example from a colleague who used the Philips model to prove the business case for funding simulation-based central line training (Barsuk et al., 2020). Because he demonstrated the ROI of the training, leaders advocated to secure funding to maintain the simulation center. Similarly, getting funding for SOTL will require demonstration of the value for each party involved and the ROI for those making financial decisions. Okuda noted that demonstrating ROI can be challenging when there is a lot of space between the initiative or intervention and the eventual outcome and said that the closer the intervention is to patient outcomes, the easier it is to demonstrate ROI.

Nunez presented the seventh and final suggestion: exploring the creation of a national center on SOTL or institute on development of teaching and learning and SOTL in HPE. The former is a "moonshot," acknowledged

Nunez, but a model can help identify a pathway. Two organizations for IPE, the National Center for Interprofessional Practice and Education and the Global Confederation for Interprofessional Education and Collaborative Practice, were created after a years-long process that brought together key individuals and groups and leveraged existing resources. Nunez shared the time line of the process for IPE as an example of how a similar journey for SOTL could proceed. An early step was creating a national framework and common language across health professions; efforts of the World Health Organization and the forum helped to accomplish this. Next, institutional or regional exemplars of IPE excellence were identified and shared as approaches that could be duplicated or scaled up, said Nunez. Core competencies were developed, and interprofessional conferences were held at the institutional, regional, national, and international levels. These efforts culminated in creating both national and global centers for IPE, which serve as change leaders to advance best practices, teaching and learning, scholarship, and advocacy across institutions as well as across nations. "What was a moonshot for IPE ultimately became a reality," said Nunez, and a similar path could be explored to achieve the moonshot for an HPE SOTL national center. She urged stakeholders to consider what is already in place for SOTL, collaborate and coalesce their efforts, and identify potential next steps to build momentum toward achieving a national or international center on SOTL.

Perspectives on HPE SOTL Priorities

Dunleavy moderated a roundtable discussion with the speakers. A number of very thought-provoking options were presented, she said, before asking each presenter to share one or two of their highest priorities.

As dean of a nursing school, Jeffries said that she intends to promote more faculty development and create a community of practice for SOTL. It would not be restricted to her school, she said, but would include participants from across HPE and other schools at Vanderbilt to break down silos and promote faculty development. Jeffries noted that, as dean, she can advocate for SOTL and fund pilot projects. She encouraged workshop attendees to go to their leaders and administrators and ask how they can help incentivize HPE SOTL.

At ASHA, said Nunez, a critical priority is to expand the visibility and value of SOTL via the research and publishing mechanisms that are available. For example, she said, the association's journal could issue calls for papers on SOTL so that readers can see that it has a place and is valued. The journal could also name section editors in HPE SOTL to demonstrate the editorial board's experience and explicitly signal what is welcome in the journal. Finally, said Nunez, the association could work closely with SOTL authors to help them with disseminating their work through other means, such as podcasts, and coordinate that with publication.

From the perspective of simulation-based education, Okuda said that a major priority will be identifying the best practices to implement, translating these into learner and patient outcomes, and disseminating this knowledge through avenues beyond traditional journals. Okuda and his colleagues in simulation education train people to advocate for its impact on learner outcomes and patient safety outcomes. In addition, they partner with like-minded organizations to advocate to broader organizations of influence (e.g., accreditors) and advocate to Congress about the importance of funding to support training and education. A health care workforce shortage is looming, especially in nursing, said Okuda, and using evidence-based simulations can make education more effective and efficient and ultimately impact patient outcomes.

As editor in chief of the open access journal *MedEdPORTAL*, Huang said that the workshop discussions reaffirmed its work. The journal has a "critical duality" of scholarship and dissemination and pushes scholars to truly add to the literature with innovative work. The workshop also affirmed that citations and other bibliometrics for journals do not tell the whole story about impact, she said. A priority would be to continue the good work of the journal in spreading the word and modeling HPE SOTL. Huang added that the workshop—and in particular the ideas of her colleagues on how to elevate and progress HPE SOTL (see Box 4-1)—influenced her identity as an educational researcher and her goal of doing work that demonstrates rigor. In her role as a faculty affairs leader talking with promotion and tenure committees, Huang said, she has a responsibility to share with others what makes SOTL unique and how to evaluate rigor, impact, and quality. Finally, on a personal level, Huang said that the workshop reminded her of the importance of bringing quality, evidence-based practices to the teaching and training of each individual learner in front of her.

CLOSING REMARKS

Huang encouraged workshop participants to "shout from the rooftops" about how HPE SOTL can improve learner outcomes and patient health. Okuda added that it is vital for people to talk about and advocate for SOTL within their own professional organizations. Many people still do not understand it, he said, and until they do, it will not make the impact that it could. Use your positions of leadership to "spread the word," said Okuda. Nunez reiterated that good work has been done in HPE SOTL but said that it is critical to build on it, strengthen it, and advocate for it. The discussions and the ideas for how to elevate and progress HPE SOTL stemming from this workshop series will help participants go back to their own institutions and play a role in advancing SOTL, she said.

Dunleavy thanked the workshop planning committee and participants and said that the energy around the workshop illustrates that people have a desire to move SOTL forward in HPE. Frost added her thanks to the planning committee and adjourned the workshop.

Appendix A

References

Allen, G. B., V. Miller, C. Nicholas, S. Hess, M. K. Cordes, J. B. Fortune, J. Blondin, T. Ashikaga, and M. Ricci. 2014. A multitiered strategy of simulation training, kit consolidation, and electronic documentation is associated with a reduction in central line-associated bloodstream infections. *American Journal of Infection Control* 42(6):643–648. https://doi.org/10.1016/j.ajic.2014.02.014.

Barsuk, J. H., W. C. McGaghie, E. R. Cohen, J. S. Balachandran, and D. B. Wayne. 2009. Use of simulation-based mastery learning to improve the quality of central venous catheter placement in a medical intensive care unit. *Journal of Hospital Medicine* 4(7):397–403. https://doi.org/10.1002/jhm.468.

Barsuk, J. H., E. R. Cohen, S. Potts, H. Demo, S. Gupta, J. Feinglass, W. C. McGaghie, and D. B. Wayne. 2014. Dissemination of a simulation-based mastery learning intervention reduces central line-associated bloodstream infections. *BMJ Quality & Safety* 23(9):749–756.

Barsuk, J. H., S. A. Barsuk, and D. B. Wayne. 2020. Return on investment from simulation-based mastery learning. In *Comprehensive healthcare simulation: Mastery learning in health professions education*, edited by W. McGaghie, J. Barsuk, and D. Wayne, 351–362. Springer Cham. https://doi.org/10.1007/978-3-030-34811-3_19.

Boyer, E. L. 1990. *Scholarship reconsidered: Priorities of the professoriate*. Carnegie Foundation for the Advancement of Teaching.

Cohen, E. R., J. Feinglass, J. H. Barsuk, C. Barnard, A. O'Donnell, W. C. McGaghie, and D. B. Wayne. 2010. Cost savings from reduced catheter-related bloodstream infection after simulation-based education for residents in a medical intensive care unit. *Simulation in Health Care* 5(2):98–102. https://doi.org/10.1097/SIH.0b013e3181bc8304.

Glassick, C. E. 2000. Boyer's expanded definitions of scholarship, the standards for assessing scholarship, and the elusiveness of the scholarship of teaching. *Academic Medicine* 75(9):877–880.

Glassick, C. E., M. T. Huber, and G. I. Maeroff. 1997. *Scholarship assessed: Evaluation of the professoriate*. John Wiley & Sons.

Hayden, J. K., R. A. Smiley, M. Alexander, S. Kardong-Edgren, and P. R. Jeffries. 2014. The NCSBN National Simulation Study: A longitudinal, randomized, controlled study replacing clinical hours with simulation in prelicensure nursing education. *Journal of Nursing Regulation* 5(2):S3–S40.

Hutchings, P., and L. S. Shulman. 1999. The scholarship of teaching: New elaborations, new developments. *Change* 31(5):10–15. https://doi.org/10.1080/00091389909604218.

Hutchings, P., M. T. Huber, and A, Ciccone. 2011. *The scholarship of teaching and learning reconsidered: Institutional integration and impact*. Jossey-Bass/Wiley.

Mokone, G. G., M. Kebaetse, J. Wright, M. B. Kebaetse, O. Makgabana-Dintwa, P. Kebaabetswe, and O. Nkomazana. 2014. Establishing a new medical school: Botswana's experience. *Academic Medicine* 89(8):S83–S87.

Shulman, L. 2001. From Minsk to Pinsk: Why a scholarship of teaching and learning? *Journal of the Scholarship of Teaching and Learning* 1(1):48–53.

Shulman, L. S., and P. Hutchings. 2004. *Teaching as community property: Essays on higher education*. Jossey-Bass.

Smith, G. C., and J. P. Pell. 2003. Parachute use to prevent death and major trauma related to gravitational challenge: Systematic review of randomised controlled trials. *BMJ (Clinical research ed.)* 327(7429):1459–1461. https://doi.org/10.1136/bmj.327.7429.1459.

Weaver, S. J., S. M. Dy, and M. A. Rosen. 2014. Team-training in health care: A narrative synthesis of the literature. *BMJ Quality & Safety* 23(5):359–372. https://doi.org/10.1136/bmjqs-2013-001848.

Wu, F. M., L. V. Rubenstein, and J. Yoon. 2018. Team functioning as a predictor of patient outcomes in early medical home implementation. *Health Care Management Review* 43(3):238–248.

Appendix B

Breakout Groups: HPE SOTL Designs

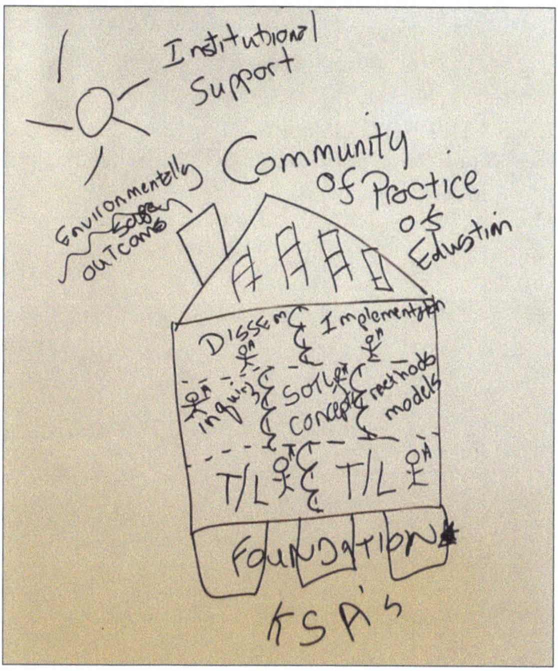

FIGURE B-1 Group 1 HPE SOTL design.
NOTES: HPE = health professional education; SOTL = Scholarship of Teaching and Learning.
SOURCE: Presented by Nancy Moreno on October 24, 2024.

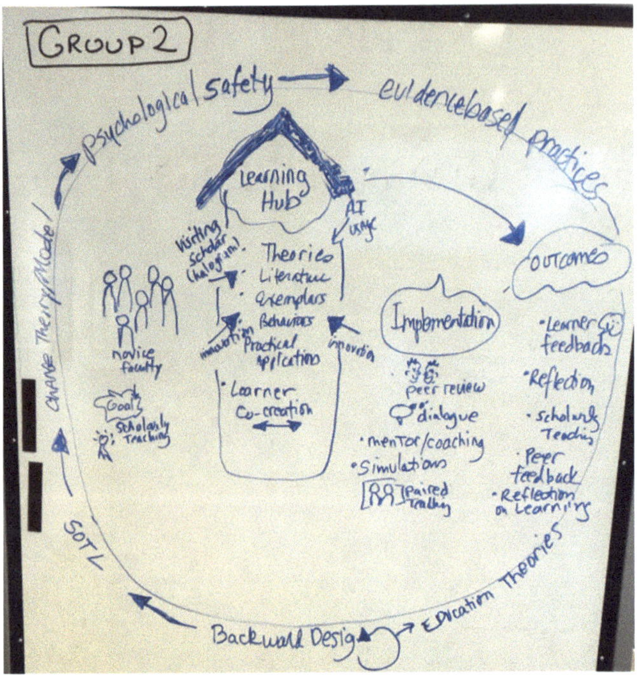

FIGURE B-2 Group 2 HPE SOTL design.
NOTE: HPE = health professional education; SOTL = Scholarship of Teaching and Learning.
SOURCE: Presented by Pamela Jeffries on October 24, 2024.

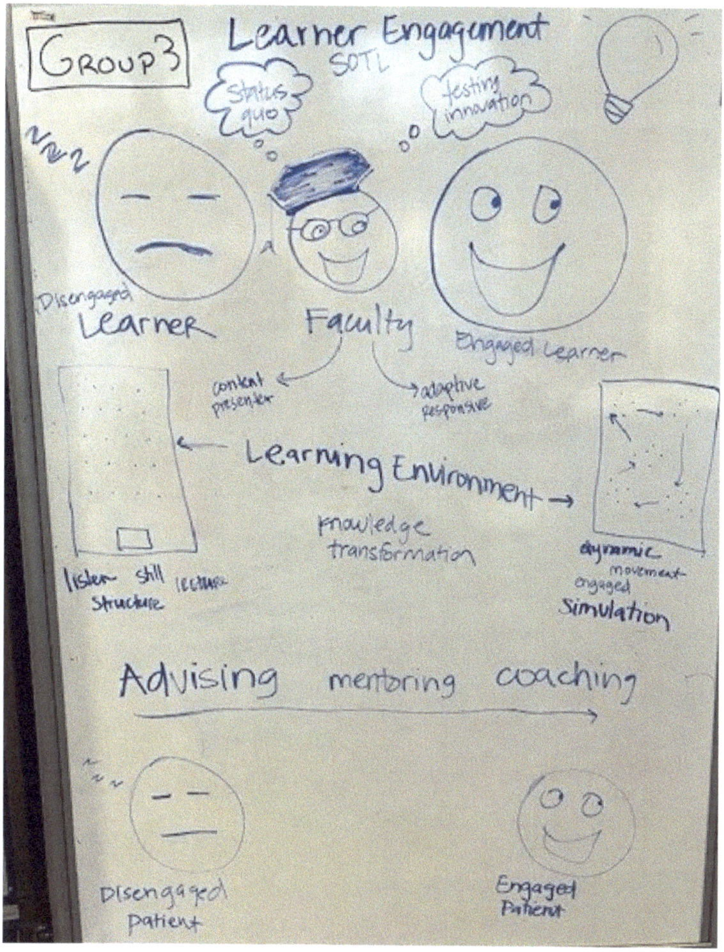

FIGURE B-3 Group 3 HPE SOTL design.
NOTE: HPE = health professional education; SOTL = Scholarship of Teaching and Learning.
SOURCE: Presented by Katie Eliot on October 24, 2024.

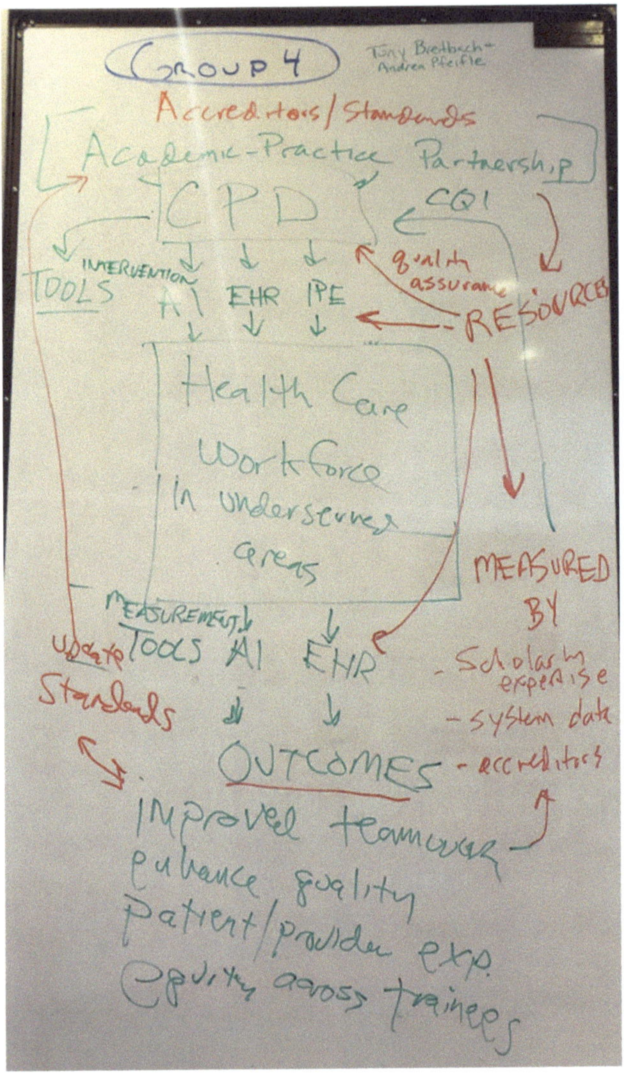

FIGURE B-4 Group 4 HPE SOTL design.
NOTE: AI = artificial intelligence; CPD = continuing professional development; CQI = continuous quality improvement; EHR = electronic health record; HPE = health professional education; IPE = interprofessional education; SOTL = Scholarship of Teaching and Learning.
SOURCE: Presented by Anthony Breitbach on October 24, 2024.

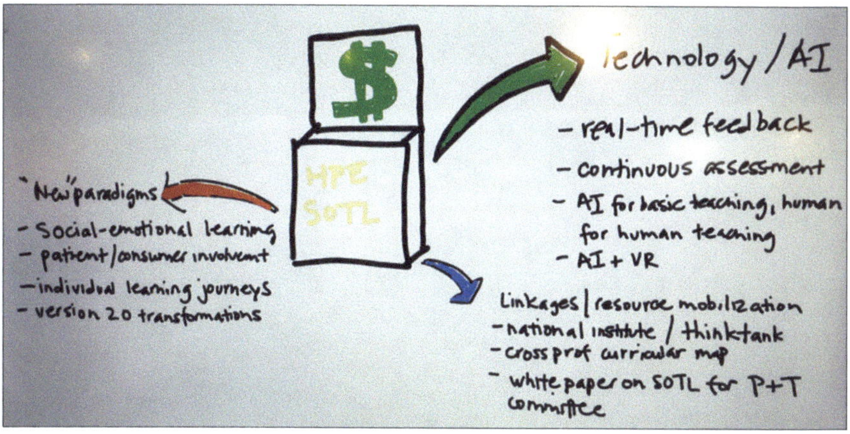

FIGURE B-5 Virtual group HPE SOTL design.
NOTE: AI = artificial intelligence; HPE = Health Professional Education; P+T = Pharmacy & Therapeutics; SOTL = Scholarship of Teaching and Learning; VR = Virtual Reality.
SOURCE: Presented by Grace Huang on October 24, 2024.

Appendix C

Workshop Agenda

SCHOLARSHIP OF TEACHING AND LEARNING IN HEALTH PROFESSIONS EDUCATION: A PREWORKSHOP

JULY 16, 2024 (VIRTUAL)
10:30 AM–11:30 AM ET

PREWORKSHOP AGENDA

Objectives

- To review the history of, and terms used in, the Scholarship of Teaching and Learning (SOTL).
- To lay a foundation for a shared understanding of SOTL and how it is applied in health professional education (HPE).
- To explore different perspectives and priorities for HPE SOTL.

10:30–10:35 am **Welcome Remarks**

- Jonathan Amiel, Senior Associate Dean, Office of Innovation in Health Professions Education Vagelos College of Physicians & Surgeons Columbia University

10:35–10:50 am	**The Scholarship of Teaching and Learning in Health Professions Education** **Moderator:** Kim Dunleavy (Planning Committee Cochair), Clinical Professor and Director of Community Engagement at the University of Florida • Gail Jensen, Vice Provost for Learning and Assessment, Dean Emerita, and Professor of Physical Therapy, Creighton University Q&A
10:50–11:25 am	**Roundtable Discussion** **Moderator:** Jonathan Amiel, Professor of Psychiatry and Senior Associate Dean for Innovation in Health Professions Education at Columbia University's Vagelos College of Physicians & Surgeons • Pamela Jeffries (Planning Committee Cochair), Dean of Vanderbilt University School of Nursing • Grace C. Huang, Dean for Faculty Affairs and Professor of Medicine, Harvard Medical School; and Hospitalist, Beth Israel Deaconess • Loretta M. Nunez, Senior Director of Academic Affairs & Research Education, American Speech-Language-Hearing Association • Yasuharu (Haru) Okuda, Executive Director and CEO, Center for Advanced Medical Learning and Simulation, University of South Florida
11:25–11:30 am	**Closing Remarks** • Kim Dunleavy (Cochair), Clinical Professor and Director of Community Engagement at the University of Florida **Adjournment**

SCHOLARSHIP OF TEACHING AND LEARNING IN HEALTH PROFESSIONS EDUCATION: A WORKSHOP

OCTOBER 23–24, 2024

WORKSHOP AGENDA
OCTOBER 23 (DAY 1), KECK K100

4:00 PM–6:00 PM ET

4:00 pm	**Introductory Session:**
	Welcome Remarks • Zohray Talib, Cochair of Global Forum on Innovation in Health Professional Education
	Framing the Workshop Objectives • Pamela Jeffries, Planning Committee Cochair • Kim Dunleavy, Planning Committee Cochair
	Background & Focus Group Efforts • Loretta Nunez, American Speech-Language-Hearing Association
4:30 pm	**The Value of Scholarship of Teaching and Learning in Health Professions Education (HPE SOTL)** **Moderator:** Gail Jensen, Creighton University • H. Carrie Chen (Virtual), Kaiser Permanente Bernard J. Tyson School of Medicine
	Audience Q&A
5:15 pm	**Synchronous & Asynchronous Activity: Instructions** **Moderator:** Grace Huang, Harvard Medical School In-person attendees remain in room to participate in table discussions. Virtual attendees: watch the Evidence-Based Videos from SOTL subject-matter experts and respond (in the survey link) to the questions below.
	NOTE: Livestream ends at 5:20 pm ET. Virtual audience, please return tomorrow when the livestream resumes at 9 am ET.

5:20 pm **Tables A–F Discussions Led by Global Forum Members**

Table A: Lynette Hamlin, Uniformed Services University

Table B: Carole Tucker, American Council of Academic Physical Therapy
a) What are the elements related to HPE SOTL that are critical and make it unique?
b) How do we connect HPE SOTL from classroom to clinic?

Table C: Ryan Scilla, Veterans Health Administration, Office of Academic Affairs

Table D: Zohray Talib, California University of Science and Medicine
a) What strategies would you use to bring HPE SOTL value back to your profession and/or across health professions?
b) How do you advocate for implementation of SOTL innovation?

Table E: Mark Merrick, Athletic Training Strategic Alliance

Table F: Lyuba Konopasek, Intealth
a) How do we promote HPE SOTL as part of professional identities?
b) How do you expand HPE SOTL training within and across health professions?

5:55 pm **Closing Remarks**
Pamela Jeffries, Planning Committee Cochair

Adjournment

WORKSHOP AGENDA

OCTOBER 24 (DAY 2), E STREET COLLABORATION ROOM

9:00 AM–12:00 PM ET

9:00 am

Day 2 Welcome Remarks
- Jody Frost, Cochair of Global Forum on Innovation in Health Professional Education
- Pamela Jeffries, Planning Committee Cochair

9:05 am

Reflections from Day 1
- Grace Huang, Harvard Medical School

9:20 am

Innovations in HPE SOTL
Moderator: Kim Dunleavy, Planning Committee Cochair
- Patricia O'Sullivan (virtual), University of California, San Francisco, School of Medicine

Q&A

10:00 am

In-Person & Virtual Activity: Instructions
Moderator: Yasuharu (Haru) Okuda, Center for Advanced Medical Learning and Simulation, University of South Florida

In-person attendees—remain in room to participate in breakout activity.
Virtual attendees—respond to the question using Slido. Livestream resumes at 11:00 am ET.

Question: How would you develop a program of the future to facilitate change using HPE SOTL? Name key elements in 300 characters or less.

10:10 am

Breakout Groups 1–4 Led by Forum Members

- Group 1: Dawn Mancuso, Association of Schools and Colleges of Optometry
- Group 2: Peter Cahn, Massachusetts General Hospital IHP
- Group 3: Katie Eliot, Academy of Nutrition and Dietetics
- Group 4: Anthony Breitbach & Andrea Pfeifle, National Academies of Practice

Background: Consider (1) the present state of health professional education and workforce; and (2) the potential for new tools, technology, and methods to change how students and health professionals learn, and how clinical and classroom educators teach.

Assignment: Illustrate an innovative HPE SOTL design that would produce a meaningful result(s) for a target population. Describe the stakeholders.

Small Group Task (In-Person Attendees): Discuss your assignment. Create a visual that summarizes your response. This can be in the form of a table, diagram, matrix, flowchart, drawing…feel free to be creative!

11:00 am	**Breakout Group Reports**
11:35 am	**Open Floor Discussion** • Kim Dunleavy, Planning Committee Cochair • Pamela Jeffries, Planning Committee Cochair
11:55 am	**Closing** • Zohray Talib, Cochair of Global Forum on Innovation in Health Professional Education

OCTOBER 31, 2024

8:30 AM–9:30 AM ET

VIRTUAL CLOSING SESSION

Purpose:
A planning committee of the National Academies of Sciences, Engineering, and Medicine organized and conducted a public workshop on the Scholarship of Teaching and Learning in Health Professions Education (HPE SOTL). The 2-day event held on October 23–24, 2024, in Washington, DC, underscored the value of HPE SOTL and explored future innovations in teaching practice leveraging HPE SOTL.

In this 1-hour public webinar, planning committee members will convene for a final closing session to provide a summary of learnings and collected feedback/input from workshop attendees for engaging more educators in SOTL.

AGENDA

8:30 am	**Welcome Remarks** • Jody Frost, IHPE Global Forum Cochair
8:35–8:45 am	**Summary of Workshop & Related Activities** • Kim Dunleavy (Cochair), Clinical Professor and Director of Community Engagement at the University of Florida; ACAPT Forum Representative
8:45–8:50 am	**Defining Mechanisms for Individual Faculty Development** • Pamela Jeffries (Cochair), Dean of Vanderbilt University School of Nursing
8:50–9:15 am	**A Call to Engage More Educators in SOTL** • Gail Jensen, Vice Provost for Learning and Assessment, Dean Emerita, and Professor of Physical Therapy, Creighton University • Loretta M. Nunez, Senior Director of Academic Affairs & Research Education, American Speech-Language-Hearing Association • Yasuharu (Haru) Okuda, Executive Director and CEO, Center for Advanced Medical Learning and Simulation, University of South Florida • Grace C. Huang, Dean for Faculty Affairs and Professor of Medicine, Harvard Medical School; and Hospitalist, Beth Israel Deaconess
9:15–9:25 am	**Roundtable Discussion: Implementing Ideas** **Moderator:** Kim Dunleavy (Cochair), Clinical Professor and Director of Community Engagement at the University of Florida & ACAPT Forum Representative **Roundtable Discussants:** • Pamela Jeffries (Cochair), Dean of Vanderbilt University School of Nursing • Loretta M. Nunez, Senior Director of Academic Affairs & Research Education, American Speech-Language-Hearing Association

- Yasuharu (Haru) Okuda, Executive Director and CEO, Center for Advanced Medical Learning and Simulation, University of South Florida
- Grace C. Huang, Dean for Faculty Affairs and Professor of Medicine, Harvard Medical School; and Hospitalist, Beth Israel Deaconess

9:25–9:30 am **Closing Remarks**
- Jody Frost, IHPE Global Forum Cochair

9:30 am **Adjournment**

Appendix D

Evidence-Based Research: Author Publications

The planning committee invited authors who represent health professional education (HPE) Scholarship of Teaching and Learning (SOTL), as outlined by Glassick (2000), to provide a short recording of their work. Their articles show Glassick's six standards that can be applied to all four forms of scholarship (clear goals, adequate preparation, appropriate methods, significant results, effective presentation, and reflective critique). Each presenter adapted these to showcase their research. Six categories believed to be **key areas** were identified through preworkshop focus groups (HPE, clinical care/clinical learning environment, simulation, interprofessional, faculty development, promotion and tenure). Videos of all presentations are available on the project webpage at https://www.nationalacademies.org/event/42801_10-2024_scholarship-of-teaching-and-learning-in-health-professions-education-a-workshop#sl-three-columns-e5a903f9-5ee2-4a45-89d1-d9476c8c1783. The following is a list of publications and a brief artificial intelligence summary[1] of their transcribed video recordings.

Heather Knight, PT, DPT, CBIS
Key Area: HPE, Physical Therapy

Knight discusses research on implementing a clinical reasoning blueprint in physical therapy education contributing to HPE SOTL by addressing the

[1] Perplexity. 2025. Perplexity (Jan 16 version) Summarizations from Evidence-Based HPE SOTL Video Transcripts. National Academies of Sciences, Engineering, and Medicine Meeting, Washington, DC [Large language model]. https://www.perplexity.ai/.

unique challenges of teaching clinical reasoning, emphasizing the need for ongoing support through communities of practice, and applying knowledge translation and implementation science to improve educator effectiveness. The study resulted in statistically significant improvements in educators' self-efficacy and a systematic organization allowing educators to select appropriate methods for different learning stages. This demonstrates how SOTL can be applied to enhance teaching practices, ultimately aiming to improve the education of future health care professionals and care quality.

Knight, H., J. Peterson, M. Reilly, J. Furze, and G. Jensen. 2024. Implementation of a clinical reasoning blueprint: Knowledge translation resources. *Clinical Teacher* 21(4):e13763. https://doi.org/10.1111/tct.13763.

Jennifer Friberg, Ed.D., CCC-SLP
Mark DeRuiter, M.B.A., Ph.D., CCC-A/SLP
Key Area: HPE, Audiology & Speech-Language Pathology

Friberg and DeRuiter discuss the establishment and implementation of an SOTL fellows program in the field of communication sciences and disorders, which includes audiology and speech-language pathology. By providing formal professional development for SOTL-curious scholars, the program aimed to address the gap between faculty members who consume SOTL research and those who actively engage in it. The program's successful approach to fostering SOTL engagement in HPE highlights the importance of mentorship, peer learning, and structured professional development in cultivating new scholars and expanding the field's research base.

Friberg, J., and M. DeRuiter. 2024. Establishing a Scholarship of Teaching and Learning fellows program. *Teaching and Learning in Communication Sciences & Disorders* 8(2). https://doi.org/10.61403/2689-6443.1347.

Alice Kim, Ph.D.
Key Area: HPE, Psychology

The study explored challenges faced by new SOTL researchers by conducting qualitative interviews with participants who had recently completed a 1-year peer-based SOTL course, including faculty from nursing and other disciplines. Kim shares important findings, such as the challenges and needs participants reported. The study showcases the value of peer-based learning, the benefit of communities of practice in advancing educational research, the importance of faculty development, and the advantage of reframing SOTL as a legitimate professional activity. The challenges and opportunities presented in this study contribute to the broader understanding of how to nurture educational scholarship in various disciplines, including HPE.

Kim, A. S. N., C. Popovic, L. Farrugia, S. A. F. Saleh, G. Maheux-Pelletier, and M. Frake-Mistak. 2020. On nurturing the emergent SOTL researcher: Responding to challenges and opportunities. *International Journal for Academic Development* 26(2):163–175. https://doi.org/10.1080/1360144X.2020.1842743.

Ana Abad-Jorge, Ed.D., M.S., RDN, FAND
Jennifer Walsh, Ph.D., RDN, M.S.
Key Area: HPE, Nutrition and Dietetics

Abad-Jorge and Walsh discuss using the Community of Inquiry framework in nutrition and dietetics education to create meaningful learning experiences by assessing classroom community and student preferences for various teaching strategies. They focused on teaching presence, social presence, and cognitive presence to support student well-being and success. The work underscores SOTL in HPE by combining theoretical frameworks, practical applications, and research to enhance student engagement and learning outcomes. *Publication in progress.*

Sarah Ginsberg, Ed.D., CCC-SLP
Key Area: Clinical Care/Clinical Learning Environment

This study compared the diagnostic reasoning processes of experienced speech-language pathologists (experts) with those of student clinicians (novices) to address a pedagogical gap between classroom teaching and clinical practice. The study drew on literature from medicine and nursing to inform its methodology and conceptual framework. Ginsberg's analysis resulted in 10 themes or heuristics that describe different reasoning strategies. These findings directly inform Ginsberg's teaching methods, allowing her to take an evidence-based approach to instruction. The study exemplifies SOTL by systematically investigating a teaching-related question, using rigorous research methods, and applying the findings to improve educational practices in speech-language pathology and related health professions.

Ginsberg, S. M., J. C. Friberg, and C. F. Visconti. 2016. Diagnostic reasoning by experienced speech-language pathologists and student clinicians. *Contemporary Issues in Communication Science and Disorders* 43:87–97. https://doi.org/10.1044/cicsd_43_S_87.

Sarah Shrader, Pharm.D., FCCP
Key Area: Interprofessional

Shrader and her team developed a new model pairing interprofessional education with practice, addressing a gap in existing approaches. They consulted literature and used established frameworks to inform their interprofessional learning and practice model, assessing both student learning and patient health outcomes. The 24-month study demonstrated improved patient diabetes control and depression screening and high satisfaction scores from students in the full model. Shrader discusses how their research exemplifies SOTL by systematically investigating an educational innovation, measuring its impact on both learners and patients, and disseminating the findings to inform HPE practice and research.

Shrader, S., S. Jernigan, N. Nazir, and J. Zaudke. 2023. Determining the impact of an interprofessional learning in practice model on learners and patients. *Journal of Interprofessional Care* 37(sup1):S67–S74. https://doi.org/10.1080/13561820.2018.1513465.

Patricia O'Sullivan, Ed.D., M.S.
Key Area: Faculty Development

O'Sullivan discusses a research study aimed to articulate a vision for faculty development in health professions that could influence organizational change to support excellence and innovation in education. The paper presents a model with four domains of faculty development: promoting individual and group development, advocating for infrastructure and resources, influencing policy and procedures, and working on organization-wide initiatives. The authors conducted a comprehensive literature review, citing over 80 references providing examples of best practices and how organizations can benefit from SOTL in faculty development to enhance professional preparation and contribute to care delivery. The approach aligns with the principles of SOTL by focusing on improving teaching and learning practices, using evidence-based methods, and contributing to the broader knowledge base in HPE.

Steinert, Y., P. S. O'Sullivan, and D. M. Irby. 2024. The role of faculty development in advancing change at the organizational level. *Academic Medicine* 99(7):716–723. https://doi.org/10.1097/ACM.0000000000005732.

Joan Friedland, M.D., M.P.H.
Nancy Moreno, Ph.D., M.S.
Key Area: Promotion & Tenure

Friedland and Moreno discuss the implementation and impact of a standards-based, peer-reviewed teaching award system at Baylor College of Medicine. Their approach recognized education as a discipline within medical schools and adapted the work of Boyer, Glassick, and Huber to medical education. By creating standardized prototypes of excellence in teaching and training reviewers to evaluate educational contributions, the award system had a comprehensive approach to recognizing various aspects of educational scholarship. An evaluation study assessing its impact found positive influence on promotion, and a high value on awards for recognition and promotion and as a measure of rigor in educational contributions. The system has led to significant changes at Baylor, demonstrating how SOTL can drive institutional transformation in HPE.

Searle, N. S., C. R. Teal, B. F. Richards, J. A. Friedland, N. L. Weigel, R. A. Hernandez, J. W. Lomax, M. Coburn, and E. A. Nelson. 2012. A standards-based, peer-reviewed teaching award to enhance a medical school's teaching environment and inform the promotions process. *Academic Medicine* 87(7):870–876. https://doi.org/10.1097/ACM.0b013e3182584130.

Jeffrey Barsuk, M.D., M.S., SFHM
Key Area: Simulation

Barsuk discusses the return on investment (ROI) from simulation-based mastery learning in health care education and the importance of building a strong business case for it as a valuable asset that ensures all learners achieve a high level of competency. Barsuk highlights three studies that demonstrated significant ROI and positive translational science outcomes, such as improved patient care practices. This emphasizes the importance of considering both the costs of interventions and the potential cost savings from improved outcomes when evaluating such programs.

Barsuk, J. H., S. A. Barsuk, and D. B. Wayne. 2020. Return on investment from simulation-based mastery learning. In *Comprehensive healthcare simulation: Mastery learning in health professions education*, edited by W. McGaghie, J. Barsuk, and D. Wayne, 351–362. Springer Cham. https://doi.org/10.1007/978-3-030-34811-3_19.

Barsuk, J. H., E. R. Cohen, J. Feinglass, S. E. Kozmic, W. C. McGaghie, D. Ganger, and D. B. Wayne. 2014. Cost savings of performing paracentesis procedures at the bedside after simulation-based education. *Simulation in Health Care* 9(5):312–318. https://doi.org/10.1097/SIH.0000000000000040.

Cohen, E. R., J. Feinglass, J. H. Barsuk, C. Barnard, A. O'Donnell, W. C. McGaghie, and D. B. Wayne. 2010. Cost savings from reduced catheter-related bloodstream infection after simulation-based education for residents in a medical intensive care unit. *Simulation in Health Care* 5(2):98–102. https://doi.org/10.1097/SIH.0b013e3181bc8304.

Schwab, B., E. N. Teitelbaum, J. H. Barsuk, N. J. Soper, and E. S. Hungness. 2018. Single-stage laparoscopic management of choledocholithiasis: An analysis after implementation of a mastery learning resident curriculum. *Surgery* 163(3):503–508. https://doi.org/10.1016/j.surg.2017.10.006.

Desiree Díaz, Ph.D., APRN, FNP-BC, CNE, CHSE-A, ANEF, FSSH, FAAN
Key Area: Simulation

Díaz discusses several key studies conducted at the University of Central Florida College of Nursing to systematically investigate the effectiveness of different educational strategies in nursing using simulation technologies. Studies include types of simulation compared with traditional clinical experiences in achieving learning goals, interruption management training, subject-matter expertise in simulation debriefing, and the feasibility of using hologram technology. The research focuses on outcomes that are relevant to clinical practice and patient safety while addressing practical challenges in nursing, such as clinical placement shortages. The work exemplifies SOTL by demonstrating a systematic, evidence-based approach to evaluating and improving teaching practices in nursing education.

Díaz, D. A., M. Anderson, P. P. Hill, S. B. Quelly, K. Clark, and M. Lynn. 2021. Comparison of clinical options: High-fidelity manikin-based and virtual simulation. *Nurse Educator* 46(3):149–153. https://doi.org/10.1097/NNE.0000000000000906.

Díaz, D. A., M. Anderson, B. Hoffman, J. V. Alegre, E. G. Simon, A. W. Gomez, A. Villamizar, and A. Wade. 2022. Feasibility of using holograms as embedded participants (EP) in simulation-based education (SBE). *International Journal of Healthcare Simulation* 1(Supplement 1):S9–S11. https://doi.org/10.54531/TDVF8904.

Díaz, D. A., L. Gonzalez, M. Anderson, V. MacKenna, E. Hoyt, and P. P. Hill. 2020. Implications of subject-matter expertise as a requirement for debriefing: A randomized control trial. *Simulation & Gaming* 51(6):770–784. https://doi.org/10.1177/1046878120945743.

Hill, P. P., D. A. Díaz, M. Anderson, S. Talbert, and C. Maraj. 2024. Remembering to resume: A randomized trial comparing combined interruption management training and simulation-based education to simulation-based education alone. *Nursing Education Perspectives* 45(1):5–11. https://doi.org/10.1097/01.NEP.0000000000001144.

Appendix E

Workshop Planning Committee Biographies

Kim Dunleavy (*Cochair*), PT, Ph.D., FNAP, is a clinical professor and director of community engagement at the University of Florida in the PT Department. She serves as the American Council of Academic Physical Therapy representative to the Global Forum. She has contributed to improve entry-level and continuing professional education in Vietnam, Haiti, Rwanda, and Suriname. Her research and Scholarship of Teaching and Learning (SOTL) have spanned interprofessional education, continuing education models, and global instructional design. Her National Institute for Occupational Safety and Health–funded grant integrates instructional design and educational methods investigating the impact and implementation of self-management to manage and limit low back pain in nursery and landscape workers. She has received multiple awards for SOTL and the Societal Impact Award from the American Physical Therapy Association. Dr. Dunleavy served on the planning committee for the National Academies of Sciences, Engineering, and Medicine/Global Forum Nonpharmacological Management of Pain and Implementation Science in health professional education workshops. Her physical therapy training from the University of Cape Town and University of Central Arkansas was followed by a Ph.D. in instructional technology from Wayne State University. She is the chair of the National Academies of Practice Physical Therapy Academy.

Pamela Jeffries (*Cochair*), Ph.D., RN, FAAN, ANEF, FSSH, is dean of Vanderbilt University School of Nursing and Valere Potter Distinguished Chair in Nursing. Dr. Jeffries is internationally known for her research and work in nursing, simulation, and health care. Throughout the academic community, she is recognized for her scholarly contributions to executive leadership, the development of innovative teaching strategies, experiential learning techniques, new pedagogies, and the delivery of content using technology. As principal investigator (PI) on grants funded by federal and state agencies and numerous national organizations, including the National League for Nursing (NLN) and National Council of State Boards of Nursing, Dr. Jeffries has provided leadership and mentorship on groundbreaking projects impacting both nursing practice and education. With the NLN, Dr. Jeffries developed the major contribution to simulation scholarship, the framework and monograph now known as the "NLN Jeffries Simulation Theory."

Jonathan Amiel, M.D., directs the Office of Professionalism and Inclusion in the Learning Environment across the New York-Presbyterian system. He trained as a medical student, resident, and chief resident at Columbia University Vagelos College of Physicians and Surgeons, where he had a 13-year tenure in the dean's office overseeing medical student education and is a professor of psychiatry. Dr. Amiel works closely with the Association of American Medical Colleges and Arnold P. Gold Foundation for Humanism in Medicine and has been invited to serve as an associate editor of *Academic Medicine*. His work focuses on competency-based education and its relationship to the development of health professionals' identities to ensure that training is intentional, just, and aligned with public health needs.

Grace C. Huang, M.D., is dean for faculty affairs and professor of medicine at Harvard Medical School (HMS). She is a nationally recognized educator with expertise in faculty development and educational scholarship; serves as the editor in chief of *MedEdPORTAL*, an online open access journal of the Association of American Medical Colleges; and is on the editorial board of *Academic Medicine*. At Beth Israel Deaconess Medical Center (BIDMC), she was vice chair for career development and mentorship in the Department of Medicine, director of academic careers and faculty development, codirector of the BIDMC Academy, and director of the Rabkin Fellowship in Medical Education. She received the 2020 Excellence in Teaching Award from the Society of Hospital Medicine, the 2021 Clifford A. Barger Excellence in Mentoring Award at HMS, and the 2022 Class Day Excellence in Clinical Instruction Award at HMS. Dr. Huang was an undergraduate at Stanford University and received her M.D. from Washington University in St. Louis, followed by internal medicine residency at Beth Israel Deaconess. She practiced medicine as a hospitalist for over 20 years.

Gail Jensen, Ph.D., M.A., is vice provost for learning and assessment, dean emerita, and professor of physical therapy. She holds a secondary appointment in the Department of Medical Humanities at Creighton University and is an adjunct professor of medicine, Center for Health Professions Education, Uniform Services University of Health Sciences, School of Medicine in Bethesda, Maryland. She is known nationally and internationally for contributions in expert practice, clinical reasoning, professional ethics, and interprofessional education. She led a National Study of Excellence and Innovation in Physical Therapist Education funded by the American Physical Therapy Association (APTA) where a key recommendation included developing more education researchers in the profession. Dr. Jensen has been instrumental in the creation and implementation of the Grant Writing and Mentorship in Education Research (GAMER) national initiative in physical therapy, now interprofessional and in its fifth year. Dr. Jensen is a Catherine Worthingham Fellow of the American Physical Therapy Association and a Distinguished Scholar Fellow in the Physical Therapy Academy of the National Academies of Practice. She is a recipient of the APTA's Rothstein Golden Pen Award, Lucy Blair Service Award, and the APTA's 2011 Mary McMillan Lecturer, and in 2022 she gave the 24th Annual Cerasoli Lecture for the Education Academy. Dr. Jensen received her Ph.D. in educational evaluation with a minor in sociology from Stanford University, holds a master's degree in physical therapy from Stanford University, and holds a bachelor's degree in education from the University of Minnesota.

Loretta M. Nunez, M.A., AuD, CCC-A/SLP, ASHA Fellow, FNAP, is senior director of academic affairs and research education for American Speech–Language–Hearing Association (ASHA). She directs activities that support academic, clinical, and research education; personnel preparation; faculty development; and higher education trends and forecasting in communication sciences disorders. Dr. Nunez leads ASHA's Envisioned Future 2025 strategic objective to advance IPE and collaborative practice. She represents ASHA on the Interprofessional Professionalism Collaborative (IPC), Interprofessional Education Collaborative (IPEC) Member Council, and National Academy of Medicine Global Forum on Innovation in HPE. She is a fellow of ASHA and distinguished fellow and practitioner of the National Academies of Practice in speech-language pathology. Dr. Nunez received the Award of Appreciation from the Council of Academic Programs in Communication Sciences and Disorders for contributions to the discipline and 2024 NAP Interprofessional Group Recognition Award for members of the IPC. She is a member of the American Interprofessional Health Collaborative and former member of its advisory board and cochair of its scholarship committee. Dr. Nunez is dually credentialed as an audiologist and speech-language pathologist and has extensive experience in clinical, academic, and association settings related to HPE.

Yasuharu (Haru) Okuda, M.D., is the executive director and chief executive officer of University of South Florida Health's Center for Advanced Medical Learning and Simulation. He also serves as associate vice president for the Office of Interprofessional Education and Practice and professor and an associate dean at the Morsani College of Medicine. Dr. Okuda was the national medical director for the Simulation Learning Education and Research Network for the Department of Veterans Affairs, where he established the national strategy and operationalized simulation-based programs at more than 160 medical facilities. He was also the founding director of the Institute for Medical Simulation and Advanced Learning for the New York City Health and Hospitals, the largest U.S. public health system. Dr. Okuda has over 20 years of experience in using emerging technologies, such as simulation and extended reality, to help train health care professionals and teams to improve patient safety and clinical care. He was given the Distinguished Educator Award by the Society for Academic Emergency Medicine Simulation Academy for creating the simulation-based training program SimWARSTM, named one of the top 25 Healthcare Leaders Under 40 by Becker's Hospital Review, selected as one of 40 Under 40 New York's Rising Stars in Business by Crain's *NY Business Magazine*, and awarded the Healthcare and Medicine Leader of the Year by *i4 Business Magazine*. In 2022, Dr. Okuda served as president of the SSH, the largest interprofessional health care association in the world. He received his B.S. in neuroscience from Brown University, M.D. from New York Medical College, and certificate in health care modeling and simulation from the Naval Postgraduate School in California. He completed a residency in emergency medicine at the Icahn School of Medicine at Mount Sinai, where he served as chief resident, and then a clinical quality fellowship from the Greater New York Hospital Association.

Appendix F

Workshop Speaker Biographies

H. Carrie Chen, M.D., Ph.D., is associate dean for medical education research and scholarship at the Kaiser Permanente Bernard J. Tyson School of Medicine. She served as the senior associate dean of assessment and educational scholarship at Georgetown University and the Abraham Rudolph Chair in Pediatric Education at UCSF. She is on the editorial board for *Advances in Health Sciences Education*, a member of the International Competency-Based Medical Education Collaborators, and core faculty for the International Course on the Ins and Outs of Entrustable Professional Activities (EPA's). Her scholarly work has focused on workplace learning and assessment, EPAs, and faculty skill development and support. Dr. Chen was a distinguished educator in the Georgetown University Medical Center Teaching Academy and an honorary member of both the Gold-Headed Cane Society and Alpha Omega Alpha at UCSF. She completed her B.S. in biochemistry at the University of California, Berkeley; M.D. and pediatric training at UCSF; M.S. in education at the University of Southern California Rossier School of Education; and Ph.D. in HPE at Utrecht University in the Netherlands.

Patricia O'Sullivan, Ed.D., M.S., directs education research for the University of California San Francisco School of Medicine after having directed both faculty development and education research for 18 years. Her research has focused on faculty development across the continuum of medical education. Additionally, she studies teaching approaches and simulations within surgical education. Dr. O'Sullivan is a fellow of the American Educational Research Association and has been acknowledged for her research and mentoring by the Society of General Internal Medicine, Association of Surgical Education, and UCSF with its Lifetime Achievement in Mentoring Award. Dr. O'Sullivan has led the Research in Medical Education Section of the Association of American Medical Colleges (AAMC), RIME Program Planning Committee, and Division for Education in the Professions of the American Educational Research Association. Her educational research studies have collaborators from medicine, nursing, and health-related professions. Dr. O'Sullivan received her B.S. in chemistry from Rosemont College, M.S. in chemistry from the University of Pennsylvania, and Ed.D. in education from the University of Houston.

Ana Abad-Jorge, Ed.D., M.S., RDN, FAND, is an associate professor at James Madison University (JMU) within the Department of Health Professions. She was the program director for the bachelor's in health sciences management at the University of Virginia. Her professional and research expertise are in pediatric clinical nutrition and nutrition support and SOTL. She has publications on pediatric and neonatal nutrition support, postgraduate nutrition education, and educational pedagogies in HPE. She has received several honors and awards, including Engaged Teacher Scholar Program and American Dietetic Association Foundation Doctoral Scholarship (2010) and Award for Excellence in Dietetics Education (2004). She is a member and fellow of the Academy of Nutrition and Dietetics, Nutrition and Dietetic Educators and Preceptors, and American Society for Parenteral and Enteral Nutrition. Dr. Abad-Jorge received her Ed.D. at Walden University in higher education and adult learning in 2013, M.S. in human nutrition from the University of Florida, and B.S. in biology from Rollins College.

Jeffrey H. Barsuk, M.D., M.S., SFHM, is the Robert Hirschtick Professor of Medicine at Northwestern University Feinberg School of Medicine, where he graduated Alpha Omega Alpha, and completed a residency in internal medicine at the McGaw Medical Center of Northwestern University at Northwestern Memorial Hospital. Dr. Barsuk completed an M.S. in clinical investigation and a certificate program in health care quality and patient safety. He serves as the executive medical director of Northwestern Simulation. For the last 20 years, Dr. Barsuk has focused his research on optimizing physician, nurse, and patient clinical performance using simulation technology. These efforts have been based on the mastery learning theoretical framework, using rigorous, validated standards for health care provider training. His team has published over 100 articles that demonstrate how this training approach improves clinical skills and patient care and outcomes while lowering health care costs. Recognized internationally as a researcher in medical education and patient safety, he has been given clinical and research excellence awards from the Society of Hospital Medicine, AAMC, and Society for Simulation in Healthcare. He has also been named in the top 10 hospitalists nationally by the American College of Physicians. He is sought internationally to lecture on how simulation-based medical education improves patient safety.

Mark DeRuiter, M.B.A., Ph.D., CCC-A/SLP, is professor and vice chair for academic affairs in the Department of Communication Science and Disorders at the University of Pittsburgh. He is the coeditor (along with Sarah Ginsberg of Eastern Michigan University) of a book that supports clinicians in their early SOTL endeavors. He is also a founding editorial board member of the journal *Teaching and Learning in Communication Sciences and Disorders*. He is a fellow of the ASHA and has received Honors of the Council of Academic Programs in Communication Sciences and Disorders. He has served on the Council for Clinical Certification in Audiology and Speech-Language Pathology, Council on Academic Accreditation in Audiology and Speech-Language Pathology, and Special Interest Group 11 (Administration and Supervision). Dr. DeRuiter chaired the 2016 Ad Hoc Committee for ASHA's Speech-Language Pathology Scope of Practice document revision and also served as the Audiology Convention cochair for the ASHA Convention in 2020 and 2022. He received ASHA's Dorothy Dreyer Award for Volunteerism. He is the cochair for speech-language pathology on ASHA's Health Care Economics Committee. He earned his Ph.D. from the University of Minnesota and M.B.A. from the Augsburg University.

Desiree A Díaz, Ph.D., APRN, FNP-BC, CNE, CHSE-A, ANEF, FSSH, FAAN, is the Endowed Enfermería Faculty Scholar, undergraduate simulation coordinator, and a professor at the University of Central Florida. She completed a postdoc from Johns Hopkins with Pamela Jeffries that focused on simulation pedagogy. Dr. Díaz creates tests and disseminates research on cutting-edge technologies, such as holograms, within the simulation community. Her strategies within simulation research pedagogy over the last decade have promoted health equity and awareness to health care disparities. She was a National Coalition Ethnic Minority Nurse Scholar and Health Leadership Fellow. She is a fellow in the American Academy of Nursing, Society for Simulation in Healthcare, and Academy of Nursing Education and immediate past president of the International Nursing Association for Clinical and Simulation Learning. Her grant funding focuses on exploring ways to eliminate health care disparities while educating the workforce. She is co-PI to a \$2.7 million HRSA grant, ENFERMERIA, which aims to empower and enhance linguistic services and culturally congruent care within the Hispanic community while encouraging advancement to graduate degrees of diverse post-baccalaureate student populations.

She is also PI to a $500,000 grant that uses simulation to examine health care disparities and the role of public health nursing, engaging multiple means to reduce health care inequity.

Jennifer Friberg, Ed.D., CCC-SLP, ASHA Fellow, is the director of scholarly teaching and Cross Endowed Chair in SOTL for the Center for Integrated Professional Development at Illinois State University. She is also a professor of CSD and served as an SOTL scholar mentor, overseeing more than 100 independent studies and two study abroad trips with students. She is the coeditor of three recent volumes: *Scholarship of Teaching and Learning in Communication Sciences and Disorders* (2024), *Going Public Reconsidered: Engaging the World Beyond Academe Through the Scholarship of Teaching and Learning* (2022), *and Evidence-Based Education in the Classroom: Examples from Clinical Disciplines* (2022). She is a founding editorial board member of the peer-reviewed journal *Teaching and Learning in Communication Sciences & Disorders*. Dr. Friberg earned her B.A. in communicative disorders from San Diego State University, M.A. in speech-language pathology from Indiana University, and Ed.D. in curriculum and instruction at Illinois State University.

Joan A. Friedland, M.D., M.P.H., is a professor of medicine at the Center for Ethics and Health Policy and Huffington Department of Education, Innovation and Technology at Baylor College of Medicine. She serves as senior deans' scholar for educational leadership. Earlier positions include attending physician and associate chief of medicine, Michael E. DeBakey Veterans Administration Hospital, and provosts' scholar for educational leadership at Baylor. She received the Barbara and Corbin J. Robertson, Jr. Presidential Award for Excellence in Education. At Baylor, she has contributed to developing two major medical school curriculum revisions, 1995 and 2023, and extensive educator development initiatives. As chair, she led the Baylor Committee for Educator Development during the design and implementation of the Norton Rose Fulbright Faculty Excellence Award. National contributions include numerous workshops and presentations at AAMC and serving as national chair of the AAMC GEA Section on Graduate Medical Education. Other work includes research and publication in physical diagnosis, quality assurance, residents' teaching skills, and faculty development and recognition. Dr. Friedland is a graduate of the University of Illinois Chicago and its College of Medicine, received an M.P.H. from the University of Texas School of Public Health, and is a 1997 Harvard Macy Scholar.

Sarah M. Ginsberg, Ed.D., CCC-SLP, ASHA Fellow, is the faculty associate director of the Faculty Development Center and a professor of CSD at

Eastern Michigan University. Dr. Ginsberg served as chair of the Special Interest Group 10: Issues in Higher Education and for several years on the coordinating committee for the vice president for academic affairs, both for ASHA. She was the founding editor of the peer-reviewed journal *Teaching and Learning in Communication Sciences & Disorders* and continues to serve on the board of directors. She is the coeditor of *Scholarship of Teaching and Learning in Communication Sciences and Disorders: Past, Present, and Future* (2024); *Clinician's Guide to Applying, Conducting, and Disseminating Clinical Education Research* (Routledge, 2024); and *Evidence-Based Education in the Classroom: Examples from Clinical Disciplines* (Routledge, 2021). Her area of research focuses on the role of communication in higher education teaching effectiveness and the intersection between university teaching support programs and SOTL. Before higher education, she worked 12 years as a clinician in hospitals with patients with neurologically based communication and swallowing disorders.

Alice Kim, Ph.D., is the chair of psychology at the University of Guelph-Humber and the founder of Teaching and Learning Research in Action. She serves as a coeditor for the *Canadian Journal for the Scholarship of Teaching and Learning* and chair elect for the Teaching of Psychology (ToP) Section of the Canadian Psychological Association. Her formal training is in experimental psychology and cognitive neuroscience of memory and learning. Her research is focused on factors that impact students' learning trajectories, including student engagement, experiential education, and application of cognitive learning principles in course design. Much of her research also explores student–faculty partnerships that foster the co-creation of learning and teaching. In 2023, she received a D2L Innovation Award in Teaching and Learning from the Society of Teaching and Learning in Higher Education. In 2024, she was the inaugural recipient of the SOTL Research Award from the Canadian Psychological Association ToP section. Dr. Kim completed both her M.A. and Ph.D. in psychology at the University of Toronto.

Heather Knight, PT, DPT, is an associate professor and director of post-professional education in the Department of Physical Therapy at Creighton University in Omaha, NE. Dr. Knight has been a board-certified clinical specialist in neurologic PT since 2011 and involved in teaching concepts related to neurologic rehabilitation in professional and post-professional education for over 10 years. Her scholarship has focused on clinical reasoning and blending evidence-based practice for neurologic rehabilitation with best practices in teaching and learning across the continuum of PT education. She is actively engaged in the APTA through the Academies of Neurologic Physical Therapy and Physical Therapy Education, which has led to an interest in knowledge translation and implementation science within these specialty areas. Dr. Knight completed post-professional training through Kaiser Permanente's neurologic PT residency program in 2010 and obtained a DPT in PT from Creighton University in 2008 and a B.S. in kinesiology from Arizona State University in 2005.

Nancy P. Moreno, Ph.D., M.Sc., is professor and inaugural chair of the Huffington Department of Education, Innovation & Technology and Thomas R. Powers Distinguished Chair of Educational Outreach at Baylor College of Medicine. She also served as associate provost of faculty development and institutional development at Baylor; she expanded faculty development and recognition opportunities and led an institution-wide effort to update institutional guidelines for faculty appointment and promotion. She is a member of the first class of faculty inductees into Baylor's Academy of Distinguished Educators; received three peer-reviewed Faculty Excellence educator awards; and was honored with Baylor's highest educational award, the Barbara and Corbin J. Robertson, Jr., Presidential Award for Excellence in Education. Nationally, she was awarded the Science Prize for Online Resources in Education by the American Association for the Advancement of Science and also elected as a fellow. Her national-level efforts focus on incorporating health and biomedical science into STEM education for all learners and developing pathways into related careers from elementary school through professional education. Dr. Moreno received her B.A. from the University of Wisconsin; M.Sc. from the Instituto National de Investigaciones sobre Recursos Bióticos, in Mexico; and Ph.D. from Rice University.

Sarah Shrader, Pharm.D., FCCP, is the senior director of academic programs and professional development at the American Association of Colleges of Pharmacy. She leads professional development programs for members to reach their potential as strong academic pharmacists. She was a tenured professor of pharmacy practice and a member of the faculty at three colleges of pharmacy over 17 years. She has focused the majority of her scholarship on SOTL and IPE. She has provided numerous presentations at national and international meetings, with several publications related to IPE and collaborative practice. Dr. Shrader earned her Pharm.D. at University of Kansas and completed her residency at the Medical University of South Carolina. She is passionate about fulfilling her professional purpose of developing, teaching, serving, and leading others to grow their potential and leave the world even better than they found it.

Jennifer Walsh, Ph.D., RDN, is an associate professor in dietetics and undergraduate dietetics program director in the Department of Health

Professions at JMU. She teaches Management in Dietetics, Counseling Skills in Dietetics, Senior Seminar, Community Nutrition, and Foundations of Nutrition Practice. She has conducted nutrition education and eating behavior research for over 15 years, primarily working with underserved communities. She has facilitated health program planning and community-based participatory research for increasing healthful food access and associated supports. Dr. Walsh became a registered dietitian in 2007 and earned her Ph.D. in food and nutrition sciences from the University of Maine in 2012. From 2011 to 2016, she worked as a statewide coordinator at the University of Florida for the Family Nutrition Program and the Expanded Food and Nutrition Education Program. She serves as the public policy coordinator for the Virginia Academy of Nutrition Dietetics.

Appendix G

Members of the Global Forum on Innovation in Health Professional Education[1]

Jody Frost, Ph.D., PT, DPT, FAPTA, FNAP (*Cochair*)
Education Consultant and Facilitator

Zohray Talib, M.D. (*Cochair*)
Executive Vice Dean for Education, Chair, Dept. of Medical Education
California University of Science and Medicine, School of Medicine

Jonathan "Yoni" Amiel, M.D.
Director
Office of Professionalism & Inclusion in the Learning Environment
New York-Presbyterian

James Arrighi, M.D.
President and CEO
Accreditation Council for Graduate Medical Education
International

Richard A. Berman, M.H.A., M.B.A.
NAM Member
Associate Vice President for Strategic Initiatives for Research & Innovation
Visiting Professor of Social Entrepreneurship
Muma College of Business, University of South Florida

[1] The National Academies of Sciences, Engineering, and Medicine's forums and roundtables do not issue, review, or approve individual documents. The responsibility for the published Proceedings of a Workshop rests with the workshop rapporteurs and the institution.

Kim Dunleavy, Ph.D., MOMT, PT, OCS
Associate Clinical Professor
Director, Professional Education and Community Engagement
Department of Physical Therapy
University of Florida
American Council of Academic Physical Therapy

Daniel M. Elchert, Ph.D., LP
Licensed Psychologist & Advanced Health Systems Research Fellow
Center for Care Delivery and Outcomes Research
Minneapolis VA Health Care System

Kathrin (Katie) Eliot, Ph.D., RD
Director, Health Professions Educator, Department of Nutritional Sciences
University of Oklahoma Health Sciences Center
Academy of Nutrition and Dietetics

Christopher Feddock, M.D., M.S., FAAP, FACP
Executive Director of CSEC
National Board of Medical Examiners

Alicia Feis, O.D., FAAO
President-Elect
Arizona College of Optometry at Midwestern University
Association of Schools and Colleges of Optometry

Sara E. Fletcher, Ph.D.
Vice President and Chief Learning Officer
Physician Assistant Education Association

Elizabeth Gatewood, DNP, FNP-C, CNE, FAANP, FAAN
Assistant Dean
Education, Technology, and Innovation
Clinical Professor
Family Health Care Nursing
University of California, San Francisco

Elizabeth (Liza) Goldblatt, Ph.D., M.P.A./H.A.
Founding Board Member
Academy of Integrative Health and Medicine/Academy of Complementary and Integrative Health

Jennifer Graebe, M.S.N., NEA-BC
Director, Accreditation in NCPD and Joint Accreditation Programs
American Nurses Credentialing Center

Lynette Hamlin, Ph.D., RN, CNM, FACNM, FAAN
Professor and Associate Dean for Faculty Affairs
USU Director, Military Women's Health Research Program
Daniel K. Inouye Graduate School of Nursing
Uniformed Services University

Kendra Harrington, M.S., PT, DPT
Director, Residency/Fellowship Accreditation
American Board of Physical Therapy Residency and Fellowship Education
American Physical Therapy Association

Karen Hauer, M.D., Ph.D.
Vice Dean for Education
Molina Family Presidential Chair in Medical Education
Professor of Medicine
University of California, San Francisco

Eric S. Holmboe, M.D.
CEO
Intealth

Lisa Howley, M.Ed., Ph.D.
Senior Director of Strategic Initiatives and Partnerships
Association of American Medical Colleges

Cheryl L. Hoying, Ph.D., RN, NEA-BC, FAONL, FACHE, FAAN
Interim CEO/COO
National League for Nursing

Barry Issenberg, M.D.
President-Elect
Society for Simulation in Healthcare

Emilia Iwu, Ph.D., M.S.N., RN, APNC, FWACN
Assistant Dean, Center for Global Health
Rutgers School of Nursing

Pamela Jeffries, Ph.D., RN, FAAN, ANEF
Dean of Nursing
Vanderbilt University

Nasreen Jessani, D.P.H., M.S.P.H.
Faculty: Senior Researcher
Stellenbosch University

Lyuba Konopasek, M.D.
Executive Director
FAIMER Division
Senior Vice President
Intealth

Kimberly Lomis, M.D.
Vice President
Medical Education Innovations
American Medical Association

Beverly Malone, Ph.D., RN, FAAN
NAM Member
Chief Executive Officer
National League for Nursing

Dawn M. Mancuso, MAM, CAE, FASAE
Executive Director
Association of Schools and Colleges of Optometry

Rance McClain, D.O.
Senior Vice President of Medical Education
American Association of Colleges of Osteopathic Medicine

Karen McDonnell, Ph.D.
Associate Professor, Milken Institute School of Public Health
George Washington University

Mark Merrick, Ph.D., ATC, FNATA
Professor and Dean
College of Health and Human Services
University of Toledo
Athletic Training Strategic Alliance

Warren Newton, M.D., M.P.H.
President and CEO Elect
American Board of Family Medicine

Loretta Nunez, M.A., Au.D., CCC-A/SLP, FNAP
ASHA Fellow
Senior Director of Academic Affairs & Research Education
American Speech-Language-Hearing Association

David O'Bryon, J.D., CAE
President
Association of Chiropractic Colleges
Immediate Past Chair
Academy of Integrative Health and Medicine and ACIH Council

John Ogunkeye, M.S.
Executive Vice President
Accreditation Council for Graduate Medical Education
Global Services

Haru Okuda, M.D., FACEP, FSSH
Executive Director & CEO
USF Health CAMLS | HPCC
Associate Vice President
USF Health IPEP
Associate Dean, Emerging Healthcare & Educational Technologies, and Professor
Morsani College of Medicine
Immediate Past President
Society for Simulation in Healthcare

Bjorg Palsdottir, M.P.A.
Training for Health Equity Network
Executive Director and Cofounder

Stephanie Petrosky, DCN, M.H.A., RDN, LDN, FAND
Associate Professor and Chair
Department of Nutrition
Director, M.S. in Nutrition Program
Nova Southeastern University
Academy of Nutrition and Dietetics

Andrea L. Pfeifle, Ed.D., PT, FNAP
Associate Vice President for Interprofessional Practice and Education
Ohio State University Wexner Medical Center
National Academies of Practice

Senthil Rajasekaran, M.D., MMHPE
Adjunct Professor
Wayne State University School of Medicine
Affiliate Professor
Harvard University TH Chan School of Public Health

Monica Sampson, Ph.D., CCC-SLP
Chief Staff Officer, Speech-Language Pathology
American Speech-Language-Hearing Association

Joanne G. Schwartzberg, M.D.
Scholar-in-Residence
Accreditation Council for Graduate Medical Education

Wendi Schweiger-Moore, Ph.D., NCC, LPC
Director, Global Capacity Building
National Board for Certified Counselors, Inc. & Affiliates

Ryan Scilla, M.D., FACP, CPE
Director
Medical and Dental Education
VA Office of Academic Affiliations

Javaid I. Sheikh, M.D., M.B.A.
Dean
Weill Cornell Medicine—Qatar
Qatar Foundation—Education City

Angela M. Simmons, Ph.D., RN, NEA-BC
Commandant, Associate Dean of Research
Uniformed Services University Graduate School of Nursing

Nancy Spector, Ph.D., RN, FAAN
Director of Nursing Education
National Council of State Boards of Nursing

Jeffery Stewart, D.D.S., M.S.
Senior Vice President for Interprofessional and Global Collaboration
American Dental Education Association

Arianne Teherani, Ph.D.
Professor of Medicine and Education Scientist
Center for Faculty Educators
School of Medicine
University of California, San Francisco

Carole Tucker, PT, Ph.D., M.S.
Chair and Professor
Physical Therapy and Rehabilitation Sciences
School of Health Professions
University of Texas Medical Branch
American Council of Academic Physical Therapy

Halaevalu (Valu) F. Ofahengaue Vakalahi, M.S.W., Ph.D.
President and CEO
Council on Social Work Education

Karen P. West, D.M.D., M.P.H.
President and CEO
American Dental Education Association

Alison J. Whelan, M.D.
Chief Medical Education Officer
Association of American Medical Colleges

Global Forum Staff

Patricia Cuff, Ed.D., M.P.H., M.S.
Forum Director and Senior Program Officer
Board on Global Health

Erika Lin Chow
Research Associate
Board on Global Health

Julie Pavlin, M.D., Ph.D., M.P.H.
Board Director
Board on Global Health